Texts and
Monographs
in Physics

W. Beiglböck
series editor

C. Truesdell and S. Bharatha

The Concepts and Logic of
Classical Thermodynamics
as a Theory of Heat Engines

Rigorously Constructed upon the Foundation
Laid by S. Carnot and F. Reech

Springer-Verlag
New York Berlin Heidelberg
London Paris Tokyo

Clifford Ambrose Truesdell, III
The Johns Hopkins University
Baltimore, Maryland 21218
USA

Subramanyam Bharatha
Research Department
Esso Resources Canada Limited
Calgary, Alberta, Canada

Wolfgang Beiglböck
Springer-Verlag
Neuenheimer Landstrasse 28–30
6900 Heidelberg, FR6
Federal Republic of Germany

Library of Congress Cataloging in Publication Data

Truesdell, Clifford Ambrose, III, 1919–
    The concepts and logic of classical thermodynamics as a theory of heat engines,
rigorously constructed upon the foundation laid by S. Carnot and F. Reech.

    (Texts and monographs in physics)
    Bibliography: p.
    Includes indices.
    1. Thermodynamics. 2. Thermodynamics—History. 3. Heat engines.
I. Bharatha, Subramanyam, 1945– joint author. II. Title: The concepts and logic of
classical thermodynamics as a theory of heat engines...
QC311.T79   536'.7   76-48115

Printed on acid-free paper.

9 8 7 6 5 4 3 2

ISBN-13: 978-3-642-81079-4   e-ISBN-13: 978-3-642-81077-0
DOI: 10.1007/978-3-642-81077-0

May this tractate be received as an expression of
respectful gratitude for the legacy of the great
French thermodynamicists

CARNOT, REECH, DUHEM.

PREFACE TO THE SECOND PRINTING

There are three major failings in the text reprinted here:

1. Only the ideal-gas scale of temperature is considered.
2. Axiom II requires $\partial K_V/\partial\theta$ to exist. Such an assumption is acceptable for any one particular scale but cannot be preserved under changes of empirical scale.
3. The historical references are not complete.

In the ten years since the first printing appeared I have removed these three deficiencies.

For the historical side in full detail down to 1854, see *The Tragicomical History of Thermodynamics 1822–1854* (Springer-Verlag, New York, 1980). For an outline of part of the work done after 1854, see "The origins of rational thermodynamics", pages 1–56 of *Rational Thermodynamics*, Second Edition (Springer-Verlag, New York, 1984); and "What did Gibbs and Carathéodory leave us about thermodynamics?", pages 101–134 of *New Perspectives in Thermodynamics*, edited by J. SERRIN (Springer-Verlag, Berlin, 1986).

The first and second shortcomings are removed in the works listed below in order of decreasing recency:

"Classical thermodynamics cleansed and cured" (1985), pages 265–291 of *Contributi del centro Linceo interdisciplinare di Scienze Matematiche e loro applicazioni* No. 76, Roma, Accademia Nazionale dei Lincei, 1986. (A corrected and revised text completed in 1987 may be had from me on request.)

"Thermodynamics for beginners", pages 82–106 of *Rational Thermodynamics*, Second Edition (Springer-Verlag, New York, 1984).

"Absolute temperatures as a consequence of Carnot's General Axiom", *Archive for History of Exact Sciences* **20** (1979), 357–380.

The first two items on this list are expository outlines. Full proofs resting upon explicit conditions of smoothness are found in the last work cited, "Absolute temperatures . . .", and in the book now reprinted. I have not published them elsewhere.

June 1988   C. Truesdell
Il Palazzetto
Baltimore

PREFACE

Mon but n'a jamais été de m'occuper des ces matières comme physicien, mais seulement comme logicien...

F. REECH, 1856

I do not think it possible to write the history of a science until that science itself shall have been understood, thanks to a clear, explicit, and decent logical structure. The exuberance of dim, involute, and undisciplined historical essays upon classical thermodynamics reflects the confusion of the theory itself. Thermodynamics, despite its long history, has never had the benefit of a magisterial synthesis like that which EULER gave to hydrodynamics in 1757 or that which MAXWELL gave to electromagnetism in 1873; the expositions in the works of discovery in thermodynamics stand a pole apart from the pellucid directness of the notes in which CAUCHY presented his creation and development of the theory of elasticity from 1822 to 1845. Thermodynamics was born in obscurity and disorder, not to say confusion, and there the common presentations of it have remained.

With this tractate I aim to provide a simple logical structure for the classical thermodynamics of homogeneous fluid bodies. Like any logical structure, it is only one of many possible ones. I think it is as simple and pretty as can be.

My approach to thermodynamics here is altogether different from those in my earlier treatments. In Chapter EII of *The Classical Field Theories* (1960) Mr. TOUPIN & I took entropy as a primitive concept and developed the formal structure of equations of state with an arbitrary number of independent variables. In my *Thermodynamics for Beginners* (1966), and later in my *Rational Thermodynamics* (1969), I took a concept like entropy as primitive and, following a line of argument similar to COLEMAN & NOLL's for thermoelasticity, set up for "irreversible" processes a structure in which "reversible" processes are included as a degenerate if important special case.

Here, in contrast, I take as primitive *just those quantities which every pioneer took as primitive*, namely:

$$t \qquad \text{time}$$

and, associated at the time t with a fluid body of given mass:

$V(t)$ volume [$V(t) > 0$]

$\theta(t)$ temperature [by convention restricted to the range $0 < \theta(t) < \infty$]

$p(t)$ pressure

$Q(t)$ heating

—these and no other. As CARNOT wrote[1] in 1824,

> I regard it useless to explain here what is...quantity of heat...or to describe how to measure [it] with the calorimeter. Neither will I explain what is meant by latent heat, degree of temperature, specific heat, *etc.* The reader should be grown familiar with these terms through study of the elementary treatises of physics or chemistry.

The reader seeking a proof of the existence of an absolute temperature on the basis of ideas about thermal equilibrium, heat, and work will not find it here. Every such "proof" I have seen seems to me to be a blend of metaphysics with vicious circles, more or less elaborately dressed up according to taste. I do not claim that no real theory of this kind can ever be found,[2] but I do hold to the opinion that thermodynamics has been castrated by thermodynamicists' morbid and manic insistence that progress be forbidden until the undefinable shall have been defined.

Moreover, I deal only with classical thermodynamics, to which the term "reversible" was attached long after the theory was completed, and only with two independent variables, which I choose to interpret as being the volume and the temperature of a homogeneous fluid body. Finally, I employ no

1. *Réflexions sur la Puissance Motrice du Feu et sur les Machines propres à Développer cette Puissance*, Paris, Bachelier, 1824, several times reprinted and translated. The passage quoted here occurs on p. 15 of the first edition. The translation is mine. The quotations on the following pages are likewise given in my translation, and the page numbers cited are likewise from the first edition. All published translations I have seen are defective at crucial points.

2. After I had fashioned the theory presented in this tractate and derived the main results from it, Mr. SERRIN sent me notes in which he does construct a mathematical theory of temperature. Of course his work is above reproach. It must not be confused with the physicists' attempt to extract a concept of temperature from ideas about thermal equilibrium, heat, and work *alone*, irrespective of constitutive relations. Rather, he introduces axiomatically an *abstract hotness manifold*, and by use not only of axioms about heat, work, and thermal interaction of materials but also of a special class of constitutive relations, expressed in terms of an empirical temperature, he proves that on the manifold there is a special scale which deserves the name "absolute". My treatment here, on the other hand, like those of CARNOT, KELVIN, CLAUSIUS (1850), and REECH, assumes from the outset the existence not only of a hotness manifold but also of a scale of this kind, motivated now as it was in the days of the pioneers by common experience with many gases, and thus does not call directly upon pairs of bodies.

Mr. SERRIN's theory and mine have some features in common and some differences. As his work is not yet published, and as he informs me his final treatment will differ in many ways from that in the notes he sent me, it would be premature to make comparisons and contrasts now.

mathematical tools that were not already available in the 1820s, in practice if not always in rigorous detail,[3] and every thermodynamic concept I use was introduced by 1854 at the latest. The entire theory is based on the systematic, elementary, and rigorous use of Carnot cycles. The central axiom is CARNOT's own: *The motive power of a Carnot cycle is a function of its operating temperatures θ^+ and θ^- and of the heat absorbed at the higher temperature* (Chapter 8).

That this tractate is a long one, results from its triple scope:

1. *Conceptual:* for those already expert in thermodynamics, to show how all the concepts of the traditional, elementary theory can be derived from simple and natural assumptions about heat engines, developed by simple and rigorous mathematics, with no "physical" arguments and no appeal to metaphysics.
2. *Pro-Historical:* for those who would study the pioneer researches, by logical analysis to reveal the features of principle *common* to CARNOT's thermodynamics and CLAUSIUS', and to discern the irreconcilable *differences* of principle between them, never before carded of metaphysics and hollow rhetoric.
3. *Paedagogical:* for those who wish to learn a clean elementary thermodynamics so as to teach it to beginners.

A reader seeking but one of these ends will find this work too long and must set parts of it aside, yet I hope for some who will put value on all three, and I know that had I chosen to write three separate memoirs, one for each purpose, their summed length would have exceeded this one's. Moreover, had I been content to follow the tradition of thermodynamics in presenting merely manipulative "proofs" that presume the adiabats of all bodies to be similar to those of an ideal gas with constant ratio of specific heats, and then take it for granted that all functions occurring are smooth and invertible at pleasure, a dozen pages would have sufficed to obtain all the formulae below. I have chosen instead to respect and face the facts that thermodynamics should be broad enough to comprehend. In some examples drawn from nature the adiabats are wholly different in kind from those of an ideal gas, certain key functions are not invertible even locally at certain states, such states cannot be inclosed by any Carnot cycle, and Carnot cycles corresponding to large differences of temperature generally fail to exist. Another factor has drawn

---

3. That my statements sometimes employ *terms* which were not precisely defined or even not yet in use in the early days of thermodynamics, should not lay me open to the charge of "present-mindedness". As an author today I cannot honestly use an obsolete and vague terminology whose unspecified connotations are long since forgotten; neither could I honestly claim that the greater precision in regard to continuity, differentiability, *etc.* which was achieved in the half century between 1850 and 1900 is essential to the clean elementary thermodynamics, missed altogether by the pioneers, which I construct in this tractate.

For example, CAUCHY in stating for the differential equation $dy/dx = f(x, y)$ his existence theorem, cited below in Footnote 1 to Chapter 4, laid down as his assumption only that $f(x, y)$ and $\partial f(x, y)/\partial y$ not be infinite at (x_0, y_0), but his proof makes it abundantly clear that he assumed both those functions to be continuous in a region including (x_0, y_0).

out the text. Namely, not only have I taken care to phrase the assumptions and theorems so as to allow the possibility, necessary if we are to represent fluids which may freeze or boil, that the independent variables be restricted to some bounded region, but also, in contrast both with the pioneer researches and also with every textbook of thermodynamics I have seen, I take explicit account of the possibility that the fluid may have an isobaric maximum density within a limited range of pressures, and the final results are so stated and proved as to be valid for such a fluid as well as for one of the common kind.

In the end I have presented the theory in such a way that this text could be made (and indeed it already has been made) the basis of a first course in thermodynamics for gifted and thoughtful undergraduates, with the proviso, nowadays difficult of fulfillment, that they master the elements of differential and integral calculus, not merely its lingo. For this reason I have included detailed proofs of propositions which to physicists and engineers may seem so obvious as to need no proof, to mathematicians so simple that anyone can prove them. I ruefully confess that in some cases I have not found these proofs so easy as I should wish.

In my long essay, *The Tragicomical History of Classical Thermodynamics, 1822–1854*, now being completed and revised for the press, I have traced the confused and tortuous early history of this congenitally crippled science, author by author and paper by paper. In contrast, this tractate does not present the history of thermodynamics, because even for the parts of the subject that are classical the order of ideas here is not the historical one, nor did any one early author introduce and develop all of them, nor are to be found in any of the works of discovery proofs so easy and precise as mine. Following a tradition long honored in mathematical writing, I name axioms, definitions, and propositions after the early authors who presented them, or main special cases of them, in essence though by no means in detail, and, so as to emphasize this tradition, I do not cite the original works. The long memoir of F. REECH[4] is an exception, for the research I present here began

---

4. F. REECH, "Théorie générale des effets dynamiques de la chaleur", *Journal de Mathématiques Pures et Appliquées* 18 (1853), 357–568. The reader will recall that CLAUSIUS introduced and developed the concept of internal energy of an ideal gas in 1850, that he did not publish his treatment of more general equations of state until 1854, and that only in that year did he introduce the concept of entropy or distinguish between reversible and irreversible processes. REECH published a short note in 1851, stating two of the major theorems from an extensive manuscript he had deposited with the French Academy. In the long memoir he published in 1853, presumably a revision of the deposited manuscript, he mentions once or twice some of the work of KELVIN, RANKINE, and CLAUSIUS up to 1850, but his approach and treatment are entirely different from theirs.

FERDINAND REECH was born in 1805 at Lampertsloch, Alsace. He studied at the École Polytechnique and became director of testing and development of marine steam engines at the French naval port of Lorient. Later he became director of the École d'Applications du Génie Maritime. He published a book on engines in 1844 and a general course of mechanics in 1852. When REGNAULT, perhaps influenced by JOULE, began to doubt that heat was an indestructible substance, REECH undertook a logical examination of the whole matter, starting from the most general form of CARNOT's axioms, axioms CARNOT himself had applied only subject to further restrictions. REECH was ready to reject the

as an attempt to resuscitate and complete his, which is virtually unknown. Another exception is the treatise of CARNOT, since that treatise not only was REECH's starting point but also is mine, and my work may be regarded as achieving the aim REECH set himself, but in vain, namely, to vindicate at long last CARNOT's ideas, though by no means the sorry mess in which CARNOT chose to present them.

Not a history of thermodynamics, this tractate is a prolegomenon to that history: an outline of the theory such as could have been written in 1854, had thermodynamics been blessed with an EULER, a CAUCHY, or a MAXWELL. That the discoverers did not find proofs so simple as those I give here, and that they failed to draw some central conclusions which in fact do follow easily from their assumptions, have nothing to do with the subject itself or the date at which it was developed, the organization of science, the social or economic circumstances of the scientists involved or not involved in its early development, or any of the other extranea now popularly invoked so as to smother thought and truth and to exalt what might well be called "group science".

Like every branch of mathematical physics, thermodynamics reflects a broad range of common experience and the results of certain experiments— few indeed, but important. The former is so universal and the latter are so widely known as to need no further mention here.

The treatment which follows below differs radically from that found in common textbooks. It is more classical than theirs, since *it rests entirely upon ideas introduced before 1854.* In particular, I am content to use the old, intuitive concept of temperature that was taken for granted by every early author and rendered formal through the vision of an ideal gas as a thermometric substance; I refrain from remarks on the intestine nature of heat, irreversible changes, and internal "disorder"; and I refer nowhere to the con-

---

axiom that heat is conserved but not ready to adopt the axiom that heat and work are uniformly interconvertible. The researches of REGNAULT, now largely forgotten, provided major impetus to the revision of thermodynamics, for they showed that the way CARNOT's theory absolutely requires the specific heats of a gas to depend upon temperature was contrary to experimental fact. Indeed, they revealed that the specific heats of natural gases were nearly constant—a fact of major importance that REECH overlooked. It is this physical fact that suggests my Axiom V, set forth in Chapter 15 of this tractate.

REECH's work on thermodynamics passed unread and scarcely noticed. He published many other papers, mostly on thermodynamics with special reference to steam engines, and at least two more books, one of these being a treatise on the same subject. He died in 1884.

REECH's name has come down in the history of science for three reasons:

1. His concept of specific heat along a path, and the theorem expressed below by (6.10), which is an obvious corollary of formulae he did publish.
2. His ideas concerning the resistance of ships, and his scaling parameter sometimes called "the Reech number".
3. His attempt to introduce force as a primitive concept of mechanics, explained in terms of a massless but extensible wire.

How thorough is the oblivion posterity has bestowed upon his work in thermodynamics may be judged from the article upon him in the *Dictionary of Scientific Biography*. Though a full column long, it leaves his great and deeply original memoir unmentioned except through categorical dismissal of "tomes on applied thermodynamics".

cepts of thermal equilibrium or quasi-static process. Whatever their usefulness for motivation, these topics play no part in the formal structure of a theory for engineers who wish to see engines run, not creep. For example, the "quasi-static process" was barely mentioned for the first time in 1853 and was altogether foreign to the early work.[5]

One concept that played a great part in the thinking of the early theorists is perpetual motion. Indeed, not only the discoverers but also the writers of textbooks today seem to follow the maxim: When reason fails, appeal to the absurdity of perpetual motion, or deny that "something" can come from "nothing". The historical weight of such metaphysical clairvoyance cannot be denied, but of course such has no place in a finished, rational theory. The fact, and fact it is, that the alleged proofs based upon denial of perpetual motion are circular or merely declaratory,[6] though it has indeed discouraged

5. In introducing what we today call an "adiabatic process", LAPLACE called it "a sudden compression", in which he was followed by CARNOT (*Réflexions*, p. 30, footnote). REECH was the first to suggest that thermodynamic theory was appropriate to slow motions, but only because changes of kinetic energy were not taken into account by it. The work of KIRCHHOFF in 1868 showed that in a viscous, heat-conducting gas it is the slow oscillations, not the fast ones, that are approximately adiabatic.

6. CARNOT's own argument, pp. 20–21 of his *Réflexions*, serves as an example. After merely describing a Carnot cycle but not calculating its motive power or obtaining any property of it beyond its reversibility, CARNOT wrote:

But, if there were means of employing heat preferable to those we have used, that is, if it were possible by any method whatever to make caloric produce a quantity of motive power larger than we have made by our first series of operations, it would suffice to draw off a portion of this power in order to cause the caloric, by the method just indicated, to go up again from the refrigerator to the furnace and re-establish things in their original state, thereby making it possible to recommence an operation altogether like the first, and so on. That would be not only perpetual motion but also the creation of boundless motive force with no consumption of caloric or of any other agent whatever. Creation of this kind is entirely contrary to the ideas presently received, to the laws of mechanics and sound physics; it is inadmissible.

If the conclusion is "contrary to...sound physics", the argument itself, drawing as it does a specific inference regarding maximum motive power without first laying down restrictions on the competing "means of employing heat", violates sound logic. As it stands, it seems to apply, if it applies to anything, to every reversible cycle. However, by bringing CARNOT's assumptions out into the open we can reduce the passage to sense. To do so, we let $C_B^+(\mathscr{C})$ and $C_B^-(\mathscr{C})$ denote the heat absorbed and the heat emitted by a body B in undergoing a cycle $\mathscr{C}$, and we let $L_B(\mathscr{C})$ denote the work done by B in that cycle.

1. CARNOT's Construction. Let the body B_1 undergo some cycle $\mathscr{C}_1$, and let the body B_2 undergo the reverse $-\mathscr{C}_2$ of a cycle $\mathscr{C}_2$ so adjusted that $C_{B_1}^+(\mathscr{C}_1) = C_{B_2}^+(\mathscr{C}_2)$. Appeal to the reversibility of heat shows that

$$C_{B_1}^+(\mathscr{C}_1) = C_{B_2}^-(-\mathscr{C}_2). \tag{H}$$

Appeal to the reversibility of work yields

$$L_{B_1}(\mathscr{C}_1) - L_{B_2}(\mathscr{C}_2) = L_{B_1}(\mathscr{C}_1) + L_{B_2}(-\mathscr{C}_2). \tag{W}$$

This much, indeed, rests upon *nothing more than reversibility* and the assumption that B_2 in undergoing $\mathscr{C}_2$ can absorb exactly as much heat as does B_1 in undergoing $\mathscr{C}_1$.

2. Application. Although CARNOT speaks of "any method whatever", his discussion refers only to Carnot cycles $\mathscr{C}_1$ and $\mathscr{C}_2$ that both have *the same operating temperatures*, and I cannot see that any definite conclusion results unless we restrict attention to these as being the competing "means of employing heat". Then we may conceive of B_1 and B_2 as absorbing heat from and emitting heat to *exactly two other bodies*: the furnace, whose temperature is θ^+, and the refrigerator, whose temperature is θ^-. In undergoing $\mathscr{C}_1$ the

many mathematicians and engineers so much as to bring them to contemn classical thermodynamics, by no means implies any defect inherent in that theory.

---

body B_1 absorbs heat from the furnace; in undergoing $-\mathscr{C}_2$ the body B_2 emits heat to the furnace. From (H) we see that after the two cycles have been completed *the furnace has neither lost nor gained heat.* If $L_{B_1}(\mathscr{C}_1) > L_{B_2}(\mathscr{C}_2)$, it follows from (W) that *the overall result of making B_1 traverse $\mathscr{C}_1$ and B_2 traverse $-\mathscr{C}_2$ is to do positive work while the net gain of heat by the furnace is null.*

3. "Sound Physics".

(a) For CARNOT.

In CARNOT's theory $C_B^+(\mathscr{C}) = C_B^-(\mathscr{C})$ for every cycle $\mathscr{C}$ that B may undergo, so that the cycles $\mathscr{C}_1$ and $-\mathscr{C}_2$ together result in null net gain of heat not only for the furnace *but also for the refrigerator.* Thus for CARNOT $\mathscr{C}_1$ and $-\mathscr{C}_2$ together serve to do positive work yet "re-establish things in their original state." This he considers "contrary to...sound physics".

(b) For CLAUSIUS, KELVIN, *etc.*

If the caloric theory is abandoned, $C_B^-(\mathscr{C})$ need not equal $C_B^+(\mathscr{C})$. Thus we cannot conclude that the refrigerator has null net gain of heat after $\mathscr{C}_1$ and $-\mathscr{C}_2$ have been completed. Not all things are "in their original state". Accordingly, CLAUSIUS, KELVIN, and others who wished to save CARNOT's conclusion had to narrow the requirements of "sound physics". So as to forbid a greater range of possibilities as being unsound, they produced one or another prohibition stronger than CARNOT's. One commonly used today is, *in order that positive work shall have been done, some heat must have passed from a hot body to a cold body.* This statement is one of those said to prohibit a "perpetual motion of the second kind".

4. First Caveat.   While there is no difficulty in applying CARNOT's concept of "sound physics", application of "perpetual motion of the second kind" stands on shaky ground. If it is to apply to the CLAUSIUS–KELVIN theory, CARNOT's construction needs abundant glossing. The working bodies B_1 and B_2 when they undergo $\mathscr{C}_1$ and $-\mathscr{C}_2$ absorb and emit heat only when at the *same temperature* as the furnace or the refrigerator. It is here that the properties of Carnot cycles are used expressly. In such cycles the bodies B_1 and B_2 never receive heat from a hotter body; they serve only as agents for carrying heat from the furnace to the refrigerator and from the refrigerator to the furnace. The traditional inference from CARNOT's construction regards as "bodies" not just B_1 and B_2, which are examples of the bodies whose nature thermodynamics is designed to analyse; *it regards as "bodies" also the furnace and the refrigerator,* which represent the *surroundings* of the bodies to which thermodynamics otherwise refers.

5. Conclusion.   It is contrary to "sound physics" that in the cycles as above constructed we should obtain $L_{B_1}(\mathscr{C}_1) \neq L_{B_2}(\mathscr{C}_2)$. In other words, if Carnot cycles $\mathscr{C}_1$ for B_1 and $\mathscr{C}_2$ for B_2 are possible and correspond to the same operating temperatures and the same amount of heat absorbed, *they do the same amount of work.* That is, for a Carnot cycle $\mathscr{C}$

$$L_B(\mathscr{C}) = G(\theta^+, \theta^-, C^+(\mathscr{C})),$$

and the function G is the same for all bodies B.

6. Comment.   The formal structure of classical thermodynamics describes the effects of changes undergone by some single body. While it allows these effects for one body to be compared with corresponding effects for another body, it does not represent the effects associated with two bodies simultaneously or in any way conjointly. CARNOT's argument, in that it refers essentially to the *surroundings* of the bodies in question, steps outside the frame of concepts used in the formal structure. Inherently, then, it cannot be a proof *in* the thermodynamics of single bodies. On the other hand, the conclusion it claims to prove refers only to processes undergone by single bodies.

I believe the frustration critical students have experienced in trying to follow CARNOT's argument arises from their attempt to regard it as a proof, which it is not. It might, indeed, become a proof if it should be expressed in the frame of a hypertheory that envisaged the effects associated with pairs of bodies and with their surroundings.

I prefer to regard CARNOT's argument, or the modification of it by CLAUSIUS, KELVIN,

In the traditional treatments the most important denial of perpetual motion, due essentially to CARNOT himself, is invoked so as to infer that a Carnot cycle delivers the maximum work possible for given heat absorbed and given extremes of temperature. This denial is applied in connection with the working of two enemy engines, one driven backward by the other. Doubtless some mathematical theory can represent the act of coupling engines, but, as is suggested by the fact that the result asserted compares the efficiencies of many different engines, each working separately, proof of it ought not to resort to anything beyond the theory of one engine of a sufficiently general kind. More were waste when less will serve.

My purpose being to present classical thermodynamics as a rational and selfcontained science, I will not again mention perpetual motion. Rather, in Chapter 13 I will give simple proofs of the two classical propositions: CARNOT's for the caloric theory and CLAUSIUS' for the theory based on balance of energy. They follow as different special cases of an elementary estimate, apparently never before noticed, which I shall derive within the general theory of REECH. At the end of that chapter we shall see by counter-

---

etc., not as a proof but as motivation for Axioms III and IV, stated below in Chapters 8 and 15, respectively. Those axioms are clear and clean. They refer *only to the bodies with which the theory deals*, and they suffice, along with the classical premisses and one further axiom, to construct the classical thermodynamics of a single fluid body.

7. Second Caveat. "Sound physics" means different things to different persons. According to CARNOT's theory, unbounded motive power is produced by repeating a sufficient number of times any Carnot cycle, or, for that matter, any cycle that produces positive motive power. The caloric is merely carried from a higher to a lower temperature, and no caloric is consumed. Thus CARNOT's own theory, although, as we shall see below, it can be unfolded with flawless logic, to some seems open to his own reproach: "contrary to...sound physics".

8. Maximum Efficiency of Carnot Cycles. This same passage is often cited as somehow showing that a Carnot cycle achieves maximum efficiency for given extremes of temperature. I cannot see that it bears on that matter in any way. CARNOT's argument to this effect was different (*Réflexions*, pp. 22–24):

It would be just to ask..., What is the sense of the word *maximum*?...
Since any re-establishment of equilibrium in the caloric can be the cause of motive power, any re-establishment of equilibrium that occurs without producing this power should be considered a true loss. But a little reflection suffices to show that every change of temperature which is not due to a change of volume...can be nothing but a useless re-establishment of equilibrium in the caloric. The necessary condition for the maximum is thus that *in the bodies employed to effect the motive power of heat there be no change of temperature which is not due to a change of volume.* Conversely, whenever this condition is fulfilled, the maximum will be attained.

Thus it was obvious to CARNOT that the Carnot cycle was the most efficient because the working body was put in contact with the furnace or the refrigerator only when already at the same temperature, so as to avoid the "useless re-establishment of equilibrium in the caloric" that through conduction of heat would reduce two bodies of different temperature to thermal equilibrium. Again three bodies are considered so as to derive a proposition that refers only to properties of one body. No theory of thermodynamics is brought to bear. There is no proof at all, only affirmation.

The proposition itself, on the other hand, is one capable of proof or disproof, once thermodynamic axioms regarding the behavior of a single body have been laid down. Such being the case, CARNOT's affirmation becomes either superfluous or false. We can reconcile the two points of view as follows: Any thermodynamic theory that does not make Carnot cycles the most efficient should be rejected. It would not conform to "sound physics".

example that in at least one theory compatible with CARNOT's General Axiom, Carnot cycles are *not* always the most efficient for given extremes of temperature.

REECH's axioms refer to Carnot cycles. Thus in the thermodynamics based upon them, *the theory of heat engines is not an application of general principles but the source of them.* The traditional "First Law" and "Second Law" of thermodynamics, for fluid bodies susceptible only of "reversible" processes, appear here as proved corollaries of theorems on heat engines; they may be found stated in Corollaries 15.1 and 15.2 in Chapter 15.

The proofs below use a mathematical tool which the authors of textbooks, following the works of discovery, seem unwilling if not unable to apply. That tool is the calculus.

Except for some of the numbers assigned to theorems and chapters, for a few turns of phrase, and for the long footnote the foregoing lines are taken from the prologue to this work which I wrote early in 1974. The local theory was then complete,[7] but obstinate difficulties blocked my repeated attempts to extend it rigorously to the whole domain of the constitutive functions as I had resolved to do, however outlandish not only global analysis but even barest mathematical hygiene may seem in the muzzy moonlight of thermodynamics.

The difficulties were overcome by Mr. BHARATHA. He provided the essential definition of *thermodynamic part* (Definition 17 in Chapter 9), which is pretty nearly the most general kind of domain to which CARNOT's ideas can apply effectively, and also the definition of *C-process* (Definition 18 in Chapter 10), which seems to be the most general class of processes to which my Efficiency Theorem (here the First Principal Lemma, Chapter 8) can be extended. He conceived and demonstrated the lemmas on Carnot webs at the end of Chapter 7, which make the First Principal Lemma easy to understand and easy to prove. He conceived and proved also the Third Principal Lemma (Chapter 9). That lemma is the key to Theorem 7, which extends my constitutive restrictions, here stated as the Second Principal Lemma (Chapter 9), to an entire thermodynamic part. He sharpened and generalized both the statements and the proofs of the estimates of efficiency (Chapter 13). Beyond these major additions to my work as it stood in 1974, Mr. BHARATHA tightened and shortened several proofs, reordered part of the main argument, and supplied criticism and analysis of CARNOT's *Réflexions* which supplemented my own.

The tractate as it is printed here results from our collaboration in 1974–1975.

7. I presented it in various lectures in the United States and in Western Europe in the years 1972–1974, for example in the second of my lectures of 1973 at Grenoble: "Les bases axiomatiques de la thermodynamique", *Entropie* No. 63 (1975), pp. 6–11; No. 64 (1975), pp. 4–10; No. 65 (1975), pp. 4–8. See also my lecture to the section of history and paedagogy at the International Congress of Mathematicians in Vancouver in 1974, published in Volume 2 of its *Proceedings*, 1975, pp. 577–586: "How to understand and teach the logical structure and the history of classical thermodynamics".

During the long period when we were preparing this tractate our work was supported, at one time or another, by two grants from the U.S. National Science Foundation to The Johns Hopkins University: one for research on rational thermomechanics, the other for studies of the early history of thermodynamics. The latter grant required that I read and analyse the forgotten memoir of REECH; the other encouraged us to develop the ideas of CARNOT and REECH far beyond their historical context.

I am indebted to Messrs. BALL, DAY, ERICKSEN, MÜLLER, and SERRIN for criticism of parts of the work as it progressed in 1972-1974, and to Mr. S. WINTERS for help in locating information on the "anomalous behavior" of water.

Mr. BHARATHA's work in completing the text was supported by the National Research Council of Canada and McMaster University, with the encouragement of Professor M. LEVINSON.

November, 1977   C. Truesdell
"Il Palazzetto", Baltimore

CONTENTS

Synopsis   xxi

Part I. CALORIMETRY   1

CHAPTER

1 Notations. Constitutive Domain. Heat and Work. Processes. Paths.
   Cycles.   3
   (Definitions 1–8)

2 Thermal Equation of State. Examples. First Reversal Theorem.   8
   (Axiom I, Definitions 9 and 10, Theorem 1)

3 The Doctrine of Latent and Specific Heats. Second Reversal
   Theorem. Fundamental Theorem of Calorimetry.   20
   (Axiom II, Theorems 2 and 3)

4 Adiabatic Processes. Laplace's Theorem.   26
   (Definitions 11 and 12 (*Ordinary and Neutral Points*), Theorem 4)

5 Examples.   32

6 Emission–Absorption Estimates.   38
   (Definition 13 (*Pro-Entropy*), Theorem 5)

Part II. CARNOT'S GENERAL AXIOM   45

7 Carnot Processes. Carnot Cycles and Webs. Lemmas regarding Heat
   and Work for Cycles in a Carnot Web.   47
   (Definitions 14 (*Carnot Cycle*), 15 (*Ordinary Carnot Cycle*), and 16
   (*Carnot Web*))

8 Carnot's General Axiom. Local Theory: Reech's First Theorem, First Principal Lemma. 57
    (Axiom III, Theorem 6, First Principal Lemma (*Efficiency Theorem for Carnot Cycles in a Web*))

9 Basic Constitutive Restrictions in a Thermodynamic Part and in the Normal Set. Pro-Entropy and Internal Pro-Energy. 66
    (Second Principal Lemma (*Local Basic Constitutive Restrictions*), Third Principal Lemma (*Escape from the Web*), Definition 17 (*Thermodynamic Part*), Theorem 7 (*Basic Constitutive Restrictions in $\mathscr{D}_{th}$*), Theorem 8 (*Pro-Entropy and Internal Pro-Energy*), Definition 17_{bis} (*Normal Set*), Theorem 7_{ext} (*Basic Constitutive Restrictions in $\mathscr{D}_n$*))

10 Cyclic Processes in $\mathscr{D}_{th}$ and $\mathscr{D}_n$. C-Processes. Efficiency Theorem and Emission–Absorption Theorem for C-Processes. Completeness Theorems. 81
    (Definition 18 (*C-Process*), Theorem 9 (*Emission–Absorption Theorem for C-Processes*), Theorem 10 (*Efficiency Theorem for C-Processes in $\mathscr{D}_{th}$*), Theorem 11 (*Completeness of Constitutive Restrictions and Energy-Entropy Relations in $\mathscr{D}_{th}$*), Theorems 9_{bis} and 10_{bis} (*Emission–Absorption and Efficiency Theorems in $\mathscr{D}_n$*), Corollary 11.2_{ext} (*Equivalence of Constitutive Restrictions and* CARNOT's *General Axiom*))

11 Properties of Ideal Gases and Van der Waals Fluids. 100

12 Relation of Motors to Refrigerators. 109
    (Theorem 12)

13 Estimates of the Efficiency of a Body undergoing a Cyclic Process. 112
    (Theorem 13)

14 The Fourth Theorem of Reech: Existence of Four Thermodynamic Potentials. 119
    (Definitions 19 and 20, Theorem 14)

Part III. UNIVERSAL EFFICIENCY OF ORDINARY CARNOT CYCLES

127

15 Universal Efficiency of Ordinary Carnot Cycles Compatible with the Existence of an Ideal Gas with Constant Specific Heats. Proof of the "First Law" and "Second Law" of Thermodynamics. 129
    (Axioms IV (*Universal Efficiency*) and V (*Ideal Gas Thermometer*), Theorem 15)

    Appendix: A System of Axioms for the Thermodynamics of Clausius. 137

16 Invariance of the Carnot Function under Change of the Unit of Temperature. Alternative to Axiom V. 140
    (Axiom Vα, Theorem 16, Axiom Vβ, Theorem 17)

EPILOGUE

145

17 Axioms for Energy and Entropy. 147

Index of frequently used symbols 152
Index of frequently used terms 153
Index of references to historical origins of thermodynamics 154
Index of references to standard presentations of thermodynamics 154

SYNOPSIS

Part I introduces the *constitutive domain* $\mathscr{D}$, the part of the volume-tempera-ture quadrant over which the constitutive functions of a body of fluid are defined. The *constitutive* functions of the body are its *pressure function* ϖ, its *latent heat* Λ_V with respect to volume, and its *specific heat* K_V at constant volume. In terms of them are defined the *heat absorbed* C^+, the *heat emitted* C^-, and the *work L* done by a body in a *process* over an interval of time. Axiom I posits the existence of a thermal equation of state. Axiom II expresses the classical "Doctrine of Latent and Specific Heats". Theorems 1 and 2 specify the *reversibility* of all processes that is implied by Axioms I and II. Part I includes also the definition of an *adiabatic process* (Definition 11) and the classic theorem of LAPLACE concerning such a process (Theorem 4).

Part II begins with the definition of a *Carnot cycle* (Definition 14): a cycle in which the body absorbs heat at some one temperature only, say θ^+, emits heat only at some lesser temperature, say θ^-, and otherwise undergoes an adiabatic process. Such a cycle is *ordinary* if it is simple and if Λ_V is of one sign on it and within it (Definition 15). A *Carnot web* (Definition 16) is the collection of Carnot cycles made by subdividing repeatedly a given ordinary Carnot cycle. These webs are the principal tool for analysis of relations between heat and work. Axiom I and the Doctrine of Latent and Specific Heats provide a simple way to calculate the work done and the heat absorbed by a cycle of a web. Both of these quantities depend upon the adiabats making up parts of the cycle in question.

CARNOT's General Axiom is stated as Axiom III: For every ordinary Carnot cycle $\mathscr{C}$,

$$L(\mathscr{C}) = G(\theta^+, \theta^-, C^+(\mathscr{C})) > 0. \qquad (8.1)$$

By applying Theorems 1 and 2 we show that G is linear in its third argument if $\mathscr{C}$ is any cycle of the given Carnot web:

$$G(x, y, C^+(\mathscr{C})) = F(x, y)C^+(\mathscr{C}). \tag{8.4}$$

This is REECH's First Theorem, here stated as Theorem 6. The properties of Carnot webs imply that F can be expressed as follows in terms of a positive function h and an increasing function g:

$$F(x, y) = \frac{g(x) - g(y)}{h(x)}. \tag{8.8}$$

This is the First Principal Lemma. From it we show that

$$C^-(\mathscr{C}) = \frac{h(\theta^-)}{h(\theta^+)} C^+(\mathscr{C}). \tag{8.15}$$

Using these results, we obtain the Second Principal Lemma:

$$\frac{\partial}{\partial\theta}\left(\frac{\Lambda_V}{h}\right) - \frac{\partial}{\partial V}\left(\frac{K_V}{h}\right) = 0, \tag{9.1}$$

$$\frac{g'}{h}\Lambda_V = \frac{\partial\varpi}{\partial\theta}. \tag{9.2}$$

These are the *basic constitutive restrictions* connecting the constitutive functions ϖ, Λ_V, and K_V with the functions g and h.

The results so far are proved for a given Carnot web. The functions g and h are associated with that web. A different web will generally give rise to different functions g and h. The Third Principal Lemma shows that if P and Q are ordinary points of $\mathscr{D}$ on the same isotherm, then

$$\frac{\dfrac{\partial\varpi}{\partial\theta}}{\Lambda_V}\ \text{at}\ P = \frac{\dfrac{\partial\varpi}{\partial\theta}}{\Lambda_V}\ \text{at}\ Q, \tag{9.8}$$

$$\frac{\dfrac{\partial\Lambda_V}{\partial\theta} - \dfrac{\partial K_V}{\partial V}}{\Lambda_V}\ \text{at}\ P = \frac{\dfrac{\partial\Lambda_V}{\partial\theta} - \dfrac{\partial K_V}{\partial V}}{\Lambda_V}\ \text{at}\ Q. \tag{9.9}$$

This lemma, which refers to points not necessarily on Carnot cycles of the same web, provides the tool for freeing the analysis altogether from use of webs, providing we remain within a *thermodynamic part* $\mathscr{D}_{\text{th}}$. Such a part, according to Definition 17, is a simply connected, open subset of $\mathscr{D}$, every one of whose points is the limit of a sequence of points at which $\Lambda_V \neq 0$; moreover, if an isotherm intersects $\mathscr{D}_{\text{th}}$ at all, it does so at at least one point where $\Lambda_V \neq 0$. The theorems that follow are proved for $\mathscr{D}_{\text{th}}$.

Theorem 7 extends the functions g and h to the whole range of temperatures in $\mathscr{D}_{\text{th}}$ and extends the relations (9.1) and (9.2) to all of $\mathscr{D}_{\text{th}}$. Theorem 8 uses the relations provided by Theorem 7 to demonstrate the existence in $\mathscr{D}_{\text{th}}$ of a *pro-entropy* H_h and an *internal pro-energy* $E_{g,h}$ such that

$$\Lambda_V = h\frac{\partial H_h}{\partial V}, \qquad K_V = h\frac{\partial H_h}{\partial \theta}, \tag{9.26}$$

$$\frac{g}{h}\Lambda_V = \varpi + \frac{\partial E_{g,h}}{\partial V}, \qquad \frac{g}{h}K_V = \frac{\partial E_{g,h}}{\partial \theta}, \tag{9.27}$$

$$Q = h\dot{H}_h, \qquad \dot{E}_{g,h} = \frac{g}{h}Q - \varpi\dot{V}. \tag{9.28}$$

At this point we generalize the concept of Carnot process. A *C-process* (Definition 18) is a cyclic process that absorbs a positive amount of heat at some one temperature θ^+, emits a positive amount of heat at some one temperature θ^-, and otherwise is adiabatic. A *C*-process for which $\theta^- < \theta^+$ is a Carnot process, and conversely. We show that (8.15), (8.4), and (8.8) hold for all *C*-processes in $\mathscr{D}_{\text{th}}$ (Theorems 9 and 10). These theorems enable us to characterize the main branches of classical thermodynamics:
1. The caloric theory of LAPLACE and CARNOT: $h = \text{const.}$
2. Uniform interconvertibility of heat and work:

$$g = Jh + \text{const.}, \qquad J = \text{const.} > 0. \tag{10.16}$$

3. Classical efficiency of Carnot cycles $(F(\theta^+, \theta^-)/J = 1 - \theta^-/\theta^+)$:

$$g = Jh + \text{const.}, \qquad J = \text{const.} > 0,$$
$$h = M\theta, \qquad M = \text{const.} > 0. \tag{10.22}$$

Theorems 7–10 are the main theorems of this tractate.

The *normal set* $\mathscr{D}_n$ of $\mathscr{D}$ is the set of all points of $\mathscr{D}$ that not only are arbitrarily near to points at which $\Lambda_V \neq 0$ but also lie upon isotherms that contain such a point. A thermodynamic part $\mathscr{D}_{\text{th}}$ is a nonempty, open, simply connected subset of $\mathscr{D}_n$. $\mathscr{D}_n$ need be neither connected nor open; it may contain infinitely many disjoint $\mathscr{D}_{\text{th}}$, the union of which need not exhaust it. Theorem 7_{ext} extends to $\mathscr{D}_n$ the basic constitutive restrictions (9.1) and (9.2).

Theorem 11 is a converse to Theorems 7 and 8; it shows that in $\mathscr{D}_{\text{th}}$ the restrictions provided by either of those theorems are *sufficient* for Axiom III to hold, Axioms I and II being assumed. Corollary 11.2_{ext} is a converse to Theorem 7_{ext}; under the same assumption, it shows that the basic constitutive restrictions (9.1) and (9.2) in $\mathscr{D}_n$ may replace Axiom III.

Two properties of ideal gases are essential to classical thermodynamics: K_V is a function of θ alone, and $J\Lambda_V = \varpi$. Neither of these follows from Axioms I, II, and III, but those axioms provide five other statements equivalent to the former and two equivalent to the latter. Demonstration of these properties of ideal gases provides the contents of Chapter 11.

Chapter 12 characterizes the caloric theory and the uniform interconvertibility of heat and work through relations between the efficiency of a motor and the coefficient of performance of that motor as run backward to act as a refrigerator.

In Chapter 13 we obtain various upper and lower bounds for the efficiency of a cyclic process in $\mathscr{D}_{th}$ and characterize processes for which those bounds are achieved (Theorem 13). If $[g - g(\theta_{min})]/h$ is an increasing function, the maximum efficiency for given extremes of temperature is achieved by Carnot cycles and by them only. Such is the case both in the caloric theory and in theories according to which heat and work are uniformly interconvertible.

Following and making more specific the ideas of REECH, in Chapter 14 we introduce the *free energy* Φ, the *enthalpy* X, and the *free enthalpy* Z in the general framework derived from Axioms I, II, and III. We show that $\Phi(V, \theta)$, $E(V, H)$, $X(p, H)$, and $Z(\theta, p)$ are thermodynamic potentials (Theorem 14). The functions g and h remain general so far.

Part III reduces the general theory to the classical thermodynamics of CLAUSIUS. It does so by imposing two further axioms. Axiom IV, due to CARNOT, asserts that one and the same function G in Axiom III applies to all bodies. That is, the efficiency of ordinary Carnot cycles is *universal*. Thus if we can evaluate G at all its arguments for some one body, we have it for all bodies. Axiom V provides a key example, the *ideal gas with constant specific heats in the whole quadrant*. Axiom V has the effect of excluding CARNOT's theory; it turns out to restrict the general structure so as to make it identical with CLAUSIUS'. Indeed, by use of the properties of ideal gases we prove that (10.22) must hold for all bodies (Theorem 15). The "First Law of Thermodynamics" is then proved as a part of Corollary 15.1; the "Second Law of Thermodynamics", as a part of Corollary 15.2. These two corollaries include the basic identities of the classical thermodynamics of reversible processes and hence deliver all the formal structure of that theory.

In Chapter 16 we consider an alternative to Axiom V. Namely, if the efficiency of Carnot cycles is invariant under change of the unit of temperature (Axiom Vα), and if there is an ideal gas having a ratio of specific heats that is a function of θ alone (Axiom Vβ), then again classical thermodynamics follows (Theorem 17).

Of the seventeen theorems in this tractate, seven were stated, more or less, and given some semblance of a proof by some author or other before 1854. We regard the following ten theorems as being largely or entirely new: 5, 7_{ext}, 8, 9, 10, 11, 13, 15, 16, 17. Furthermore, in all previously published work on thermodynamics along classical lines that we have seen, only one[1] even attempts to establish smoothness of the functions obtained or to show that statements first established by local arguments can be extended so as to become general theorems of thermodynamics.

---

1. J. E. TREVOR, *The General Theory of Thermodynamics*, Boston *etc.*, Ginn, 1927. TREVOR assumed as axioms both of the traditional laws of thermodynamics. Some of his arguments, especially in his Chapters VII and VIII, are somewhat similar to some of ours. He attempted to specify the smoothness he presumed in thermodynamic functions and to extend merely local results. Aware of the difficulty presented by the fact that for some bodies Carnot cycles with arbitrarily large differences of operating temperatures fail to exist, he attempted to obviate it by introducing sequences of overlapping Carnot cycles.

PART I

CALORIMETRY

CHAPTER 1

Notations. Constitutive Domain. Heat and Work. Processes. Paths. Cycles.

Adopting the usage of physics, we sometimes use the same symbol to denote both a physical quantity and some function whose value it may be. Seeking to adjust the style of mathematics today to the tradition of thermodynamics, which, following the pioneer researches, avoids like the plague any mention of such details as continuity and differentiability, here we attempt to keep analytical points to a minimum consistent with common logical hygiene. Of course we eschew both differentials and subscripts to partial derivatives, those hallmarks of obscurantist thermodynamics.

The term *function* shall always mean a mapping, called in the older literature a "single-valued function". The values of all functions are real. The domains of functions are either real intervals or regions of the real plane. A function f is *increasing* if $f(x) > f(y)$ whenever $x > y$; *nondecreasing* if $f(x) \geqq f(y)$ whenever $x > y$. *Decreasing* and *nonincreasing* functions are defined similarly. The letter t denotes the *time*, and a superimposed dot denotes the derivative of a function of time alone: $\dot{f} \equiv df/dt$.

The concept of "body" will never be rendered formal in the theory. Our use of that word serves only to make present in mind an object in nature to which we might think of applying the theory, namely, one we should like to regard as being a given, fixed amount of some particular fluid material suffering homogeneous conditions, for example, a one-pound mass of pure water, all parts of which have at any one time a common density and a common temperature.

In its abstraction, the theory specifies first of all the properties *common* to all members of the class of natural bodies it is designed to represent. Beyond that, a "body" will be modeled in the theory by four *constitutive quantities*, designed to reflect the *diversity* of natural bodies. Choice of these four quantities allows the theorist to adjust the theory so as to conform, in some measure, with the behavior of observed fluid bodies. The first of these quantities is a **constitutive domain** $\mathscr{D}$:

$\mathscr{D}$ *is a nonempty, connected, open set in the positive V-θ quadrant.*

We think of $\mathscr{D}$ as representing the totality of simultaneous volumes and temperatures some natural body might have, except, possibly, for such pairs as might lie on the boundary of $\mathscr{D}$.

In the positive V-θ quadrant the lines $\theta =$ const. and $V =$ const., respectively, are conveniently denoted by their classical names: *isotherms* and *isochors*. By an *isothermal segment* we shall understand a finite or infinite interval on an isotherm, not reducing to a point. In particular, an isotherm is an isothermal segment.

Remark. Natural bodies may freeze or melt, boil or liquefy. For the purposes of the present theory, which does not take account of phase changes, different phases of a natural body may be regarded as being in fact different bodies, each with its own domain $\mathscr{D}$. Thus $\mathscr{D}$ will generally be only a part of the V-θ quadrant. In some typical applications $\mathscr{D}$ is not convex. While we require that $\mathscr{D}$ be connected, we do so not from mathematical necessity but only because the interpretation suggests we should.

Definition 1 (Heats Absorbed and Emitted). Let Q be an integrable[1] function on the interval of time $[t_1, t_2]$, $t_2 > t_1$. Then

$$C \equiv \int_{t_1}^{t_2} Q dt,$$

$$C^+ \equiv \frac{1}{2} \int_{t_1}^{t_2} (|Q| + Q) dt \geqq 0, \tag{1.1}$$

$$C^- \equiv \frac{1}{2} \int_{t_1}^{t_2} (|Q| - Q) dt \geqq 0,$$

so

$$C = C^+ - C^-. \tag{1.2}$$

1. So as to remain within the range of mathematical tools available to the pioneers of thermodynamics, by a function "integrable" on $[t_1, t_2]$ we shall always mean a piecewise continuous function, that is, a function for which there is a partition $t_1 < a_1 < a_2 < \cdots < a_n < t_2$ of the interval $[t_1, t_2]$ such that the function is continuous on the open subintervals $]t_1, a_1[,]a_1, a_2[, \ldots,]a_n, t_2[$ and can be continuously extended to the closed intervals $[t_1, a_1], [a_1, a_2], \ldots, [a_n, t_2]$. At $t_1, a_1, a_2, \ldots, a_n, t_2$ the function need not be defined. The integral of the function on $[t_1, t_2]$ is defined as the sum of the integrals of the continuous extensions on the subintervals. Each of these integrals exists as a Cauchy-integral. Their sum is independent of the choice of partition. This definition, which represents the idea of integral used in a mainly informal way by EULER and other mathematicians of the eighteenth century, is adequate for our purpose.

If $Q(t)$ is the *heating* of the given body at the time t, then the numbers C, C^+, and C^- are, respectively, the *net gain of heat*, the *heat absorbed*, and the *heat emitted* by the body in $[t_1, t_2]$.

We shall often have occasion to consider integrals like those on the right-hand sides of the second and third lines of (1.1). Because

$$\tfrac{1}{2}(|Q| + Q) = \begin{cases} Q & \text{if } Q \geq 0, \\ 0 & \text{if } Q \leq 0, \end{cases}$$

$$\tfrac{1}{2}(|Q| - Q) = \begin{cases} -Q & \text{if } Q \leq 0, \\ 0 & \text{if } Q \geq 0, \end{cases} \tag{1.3}$$

it is natural to think of C^+ as being an integral over the set of times $\mathcal{T}^+$ at which $Q > 0$, and C^- as an integral over the set of times $\mathcal{T}^-$ at which $Q < 0$. (The theory of Lebesgue integration, which of course was not known to the pioneers, makes this idea not only natural but also rigorously explicit.) Henceforth we shall use the suggestive notations defined as follows:

$$\int_{\mathcal{T}^+} fQdt \equiv \frac{1}{2} \int_{t_1}^{t_2} (|Q| + Q)fdt,$$

$$\int_{\mathcal{T}^-} fQdt \equiv -\frac{1}{2} \int_{t_1}^{t_2} (|Q| - Q)fdt, \tag{1.4}$$

f being an integrable function. In this notation

$$C^+ = \int_{\mathcal{T}^+} Qdt, \qquad C^- = -\int_{\mathcal{T}^-} Qdt. \tag{1.5}$$

Historical Comment. C^+ and C^- were introduced, in effect, by CARNOT.[2]

Definition 2 (Process). A *process* undergone by a given body is a piecewise smooth map V, θ of $[t_1, t_2]$ into $\mathcal{D}$. The process *occupies* the point $(V(t), \theta(t))$ at the time t; the process *traverses* its range.

Historical Comment. In constructing thermodynamics on the basis of processes, not small or "quasi-static" departures from equilibrium, we follow CARNOT, KELVIN, CLAUSIUS, and other pioneers. CARNOT's term for a process was "une opération".

Remark. According to this definition, a process is a pair of functions, V and θ, the values of which at the time t define a point $(V(t), \theta(t))$ in $\mathcal{D}$. The context will make clear whether the letters V and θ represent co-ordinates of a point in $\mathcal{D}$ or a pair of functions having such co-ordinates as their values.

2. *Réflexions*, p. 37, footnote: "les quantités de chaleur absorbées ou dégagées". *Cf.* also pp. 56–57.

The definition imposes the following conditions upon the two functions V and θ:

a. They are continuous in $[t_1, t_2]$.
b. There is a partition $t_1 < a_1 < a_2 < \cdots < a_n < t_2$ of the interval $[t_1, t_2]$ such that the restrictions of the functions to the subintervals $[t_1, a_1]$, $[a_1, a_2], \ldots, [a_n, t_2]$ are continuously differentiable. In particular, V and θ are differentiable except at a finite number of times in $]t_1, t_2[$.

Definition 3 (Work). Let a process have been given, and let p be an integrable function on $[t_1, t_2]$. Then

$$L \equiv \int_{t_1}^{t_2} p\dot{V}dt. \tag{1.6}$$

This integral exists because $\dot{V}$ is an integrable function. If $p(t)$ is the *pressure* exerted upon the given body at the time t, then L is the *work done* by the body subjected to the pressure p in the given process.

Definition 4. A process is *isothermal* if θ has a constant value; *isochoric* if V has a constant value; *constant* if both V and θ have constant values.

Definition 5 (Reverse of a Process). The *reverse* of a process V, θ is the pair of functions on $[t_1, t_2]$ having at the time t the values $V(t_1+t_2-t)$, $\theta(t_1+t_2-t)$. It is also a process.

Definition 6 (Simple Process). A process is *simple* if it occupies no point twice, except, possibly, that it may end at the point where it began, and if in each of the subintervals where the restrictions of V and θ are continuously differentiable, $\dot{V}$ and $\dot{\theta}$ evaluated for the restrictions do not vanish simultaneously at any time.

Definition 7 (Cyclic Process). The process V, θ is *cyclic* if $V(t_2)=V(t_1)$ and $\theta(t_2) = \theta(t_1)$.

Definition 8 (Path, Cycle). A process V, θ in $[t_1, t_2]$ is *equivalent* to the process V^*, θ^* in $[t_1^*, t_2^*]$ if there is a piecewise smooth, increasing parameter-transformation[3] ϕ from $[t_1, t_2]$ onto $[t_1^*, t_2^*]$ such that $V(t)=V^*(\phi(t))$ and $\theta(t)=\theta^*(\phi(t))$ for all t in $[t_1, t_2]$. The equivalence class constituted by all processes equivalent to a given one is called the *path* generated by the given process. All processes in a path traverse the same range. We may say also that a process *traverses* the path generated by it. Points in the common range of the processes of a path lie *on* that path. Necessarily they are points of $\mathcal{D}$.

The *reverse* of a path $\mathcal{P}$, denoted by $-\mathcal{P}$, is the equivalence class generated by the reverse of any process in $\mathcal{P}$. The equivalence class generated by a simple

3. The transformation ϕ satisfies the properties a and b in the remark after Definition 2, and its continuously differentiable restrictions have positive derivatives.

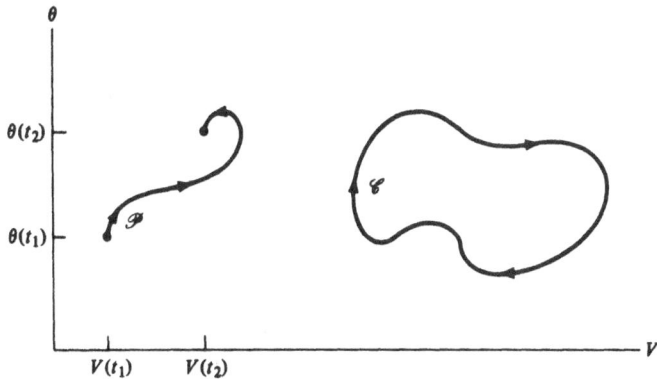

Figure 1. Simple path $\mathscr{P}$ and simple cycle $\mathscr{C}$.

process is a *simple path*; if its elements are not cyclic, it may be identified with the range of a process in it, described straight through from one end to the other, without reversals or stops, in one of the two possible senses. On a diagram such as Figure 1 the sense is conveniently denoted by arrows.

The equivalence class generated by a cyclic process is a *cycle*. A cycle is *simple* if the cyclic processes in it are simple. Necessarily, any process in a simple cycle traverses the boundary of a nonempty, bounded, simply connected, open set. Points of that open set are *included* by the simple cycle. These points need not all be points of $\mathscr{D}$. All the processes in a simple cycle traverse the same range exactly once in the same sense and without stopping. Henceforth we agree not to distinguish between simple cycles which have the same range and sense but different starting points. A simple cycle may be then identified with the range of a process in it, traversing the range exactly once in one of the two possible senses without stopping. The senses are usually denoted anthropomorphically: such as to make the inclosed region lie upon the right-hand or left-hand side of a person astride a point as it progresses. We shall say that the simple cycle is *oriented* in one or the other of these ways. Equivalently, a simple cycle may have a *clockwise* or *counterclockwise* orientation.

Let $\mathscr{P}$ be the path generated by a certain process in $[t_1, t_2]$. Taking any time $\bar{t}$ such that $t_1 < \bar{t} < t_2$, we may regard the given process as being the succession of two processes, one in $[t_1, \bar{t}]$ and the other in $[\bar{t}, t_2]$. These processes generate paths $\mathscr{P}_1$ and $\mathscr{P}_2$, respectively, the terminal point of any process in $\mathscr{P}_1$ being the initial point of any process in $\mathscr{P}_2$. We regard $\mathscr{P}$ as being *subdivided* into $\mathscr{P}_1$ and $\mathscr{P}_2$, and we write $\mathscr{P} = \mathscr{P}_1 + \mathscr{P}_2$.

Remark. The pioneers of thermodynamics used the notion of path and cycle informally, as did CAUCHY in his theory of complex integration.

CHAPTER 2

Thermal Equation of State. Examples. First Reversal Theorem.

Axiom I (Existence of a Thermal Equation of State, EULER, 1757). *The pressure p acting upon a given body is determined by the volume and temperature of that body:*

$$p = \varpi(V, \theta), \tag{2.1}$$

the domain of ϖ being $\mathscr{D}$. The function ϖ is continuous and has continuous partial derivatives $\partial\varpi/\partial V$ and $\partial\varpi/\partial\theta$; also

$$\frac{\partial\varpi}{\partial V} < 0. \tag{2.2}$$

Remark. The relation (2.1) is called the *thermal equation of state* of a particular *fluid body* described by the theory; the function ϖ is the **pressure function** of that body. The function ϖ, like the domain $\mathscr{D}$, is a constitutive quantity.

Definition 9. Any curve[1] in $\mathscr{D}$ along which ϖ has a constant value is called an *isobar* of the body.

Remark 1. Differentiation of ϖ along an isobar represented by the functions V, θ of a parameter s yields

$$\frac{\partial\varpi}{\partial V} V' + \frac{\partial\varpi}{\partial\theta} \theta' = 0. \tag{2.3}$$

---

1. By a curve we shall always mean a set of points that has a continuously differentiable parametric representation: $V = V(s)$, $\theta = \theta(s)$ for s in some finite or infinite interval. Moreover, if $V' \equiv dV/ds$ and $\theta' \equiv d\theta/ds$, we assume that V' and θ' do not simultaneously vanish for any s.

Since V' and θ' cannot vanish together for any s, from (2.2) and (2.3) it follows that the continuous function $\theta' \neq 0$ always. Hence θ is invertible, and V is a function of θ along an isobar. Further, this function is continuously differentiable and satisfies

$$\frac{dV}{d\theta} = -\frac{\dfrac{\partial \varpi}{\partial \theta}}{\dfrac{\partial \varpi}{\partial V}}. \tag{2.4}$$

Conversely, consider any solution of the ordinary differential equation (2.4) on an interval of temperatures. It is easy to verify that this solution defines a curve which is an isobar for the body with the given pressure function. Thus (2.4) is the differential equation of an isobar. Since the right-hand side of (2.4) is a continuous function of V and θ, PEANO's existence theorem shows that through every point of $\mathscr{D}$ passes an isobar, not necessarily unique. On the other hand, (2.2) and the implicit function theorem show that through each point of $\mathscr{D}$ passes an isobar unique in the following sense: Two isobars defined over the same interval of temperatures are the same. The interval of temperatures on which an isobar through a given point is defined may be extended to a maximum open interval, which may be finite or infinite. The corresponding isobar is the *maximal isobar* through the given point. An example is given a little below in connection with the Van der Waals fluid.

Of course, the pioneers of thermodynamics took the existence and uniqueness of isobars for granted.

Remark 2. The constitutive inequality (2.2) asserts that *along any isothermal segment in $\mathscr{D}$, the pressure acting upon the body is a decreasing function of the volume of that body*. This inequality has been applied in all researches on thermodynamics from the very beginning. Equations of state that do not always satisfy it are often considered, but then the part of the V-θ quadrant where it is violated is dismissed as being "unstable" or "without physical reality".

Remark 3. In the early work on thermodynamics the inequalities

$$\varpi > 0, \qquad \frac{\partial \varpi}{\partial \theta} > 0 \tag{2.5}$$

were assumed, either expressly or tacitly. They seem to represent the behavior of most fluids. The first asserts that the fluid cannot support tension. To interpret the second, we may appeal to (2.4) and by use of the fact that (2.2) holds always conclude that *lowering the temperature of a fluid body along an isobar decreases its volume*.

On the other hand, it has been known since the seventeenth century that water at atmospheric pressure experiences its maximum density at about 4°C. That is, on the isobar at 1 atm V decreases with θ to a minimum at about 4°C and then at still lower temperatures increases. This fact is commonly called "the anomalous behavior of water". From (2.2) and (2.4) we see that *along an isobar, V is an increasing function of θ in a part of $\mathcal{D}$ where the second inequality of (2.5) holds, a decreasing function of θ in a part where*

$$\frac{\partial \varpi}{\partial \theta} < 0. \tag{2.6}$$

Thus the anomalous behavior of water forbids us from imposing any general restriction upon the sign of $\partial \varpi / \partial \theta$.

Definition 10. A point of $\mathcal{D}$ where

$$\frac{\partial \varpi}{\partial \theta} = 0 \tag{2.7}$$

is a *piezotropic point*. A nonempty open set of piezotropic points is a *piezotropic part* of $\mathcal{D}$. A curve of piezotropic points is a *piezotrope*.

Remark 1. We have borrowed the term "piezotropic" from meteorology. In a piezotropic part, $\varpi = f(V)$ locally. Fluid bodies such that $\varpi = f(V)$ are often considered in classical gas dynamics. Such bodies are obviously useless as working bodies of heat engines, since the pressures exerted by them are unaffected by changes of temperature.

Remark 2. From (2.4) we see that an *isobar through a piezotropic point cuts the isotherm normally there. At all other points on an isobar, $dV/d\theta \neq 0$.* Therefore, *along an isobar the volume passes through extremes at piezotropic points, and at them only.* Thus a curve whose points are points of "anomalous behavior" for the isobars on which they lie must be a piezotrope.

Remark 3. A piezotrope is the locus of extremes of volume along isobars. Using a parametric representation, we see that along a piezotrope $p' = (\partial \varpi / \partial V)V'$, so by (2.2) it follows that *on a piezotrope p decreases when V increases.* If $\partial \varpi / \partial \theta$ and $\partial \varpi / \partial V$ have continuous partial derivatives, some further results follow. By differentiating (2.7) along a piezotrope we obtain a differential relation which its parametrization must satisfy:

$$\frac{\partial^2 \varpi}{\partial V \partial \theta} V' + \frac{\partial^2 \varpi}{\partial \theta^2} \theta' = 0. \tag{2.8}$$

The curvature of an isobar at a piezotropic point is $-(\partial^2 \varpi / \partial \theta^2)/(\partial \varpi / \partial V)$. A piezotrope on which $\partial^2 \varpi / \partial \theta^2 > 0$ is the locus of local minima of volume upon the isobars. In order for an isobar to experience an inflexion at a piezotropic point, it is necessary that $\partial^2 \varpi / \partial \theta^2 = 0$ there. If $\partial^2 \varpi / \partial V \partial \theta \neq 0$ at a piezotropic point where $\partial^2 \varpi / \partial \theta^2 = 0$, exactly one piezotrope passes through it, and (2.8) shows that there the isobar and piezotrope have vertical tangents. On the assumption that $\partial^2 \varpi / \partial \theta^2 > 0$, the sign of $d\theta/dp$ on a piezotrope is the sign of $\partial^2 \varpi / \partial V \partial \theta$, a fact that may be charitably attributed to VAN DER WAALS (1877).[2]

EXAMPLE 1. The fluid body such that

$$\varpi = R\theta/V, \tag{2.9}$$

R being a positive constant, is called the body of *ideal gas* whose *constitutive constant* is R. In this example, which was the only one familiar to the pioneers of thermodynamics, $\mathcal{D}$ may be the entire positive V-θ quadrant or a part of it. The isobars are segments of the rays through the origin. Since (2.9) contradicts (2.7), a body of ideal gas has no piezotropic points.

Remark 1. Ideal gases played a great part in the development of the early thermodynamics and have never lost their importance for it. If we grant the existence of natural fluids whose pressures, volumes, and temperatures, in the sense that experimentists use those words, obey (2.9) to within experimental error in some part of the V-θ quadrant, we may use them to

2. Since $\partial^2 \varpi / \partial \theta^2 > 0$ on the piezotrope, an argument similar to the one after (2.3), applied to (2.8), shows that along it θ is a continuously differentiable function of V satisfying

$$\frac{d\theta}{dV} = -\frac{\dfrac{\partial^2 \varpi}{\partial V \partial \theta}}{\dfrac{\partial^2 \varpi}{\partial \theta^2}}.$$

From (2.1), p is a function of V along the piezotrope, and

$$\frac{dp}{dV} = \frac{\partial \varpi}{\partial V} < 0.$$

This function is clearly invertible, and therefore θ is a function of p on the piezotrope, its derivative being given by

$$\frac{d\theta}{dp} = \frac{d\theta/dV}{dp/dV} = -\frac{\dfrac{\partial^2 \varpi}{\partial V \partial \theta}}{\dfrac{\partial \varpi}{\partial V} \cdot \dfrac{\partial^2 \varpi}{\partial \theta^2}}.$$

The statement in the text above follows from (2.2). Thermodynamicists like to express it in terms of the "coefficient of compressibility" β_1, defined as follows:

$$\beta_1 \equiv -\frac{1}{V \dfrac{\partial \varpi}{\partial V}}.$$

Since

$$\frac{\partial \beta_1}{\partial \theta} = \frac{\dfrac{\partial^2 \varpi}{\partial \theta \partial V}}{V \left(\dfrac{\partial \varpi}{\partial V} \right)^2},$$

$\partial \beta_1 / \partial \theta$ and $\partial^2 \varpi / \partial \theta \partial V$ have the same sign; if, therefore, $\partial^2 \varpi / \partial \theta^2 > 0$ on a piezotrope, then the sign of $d\theta/dp$ is the sign of $\partial \beta_1 / \partial \theta$. This is the conclusion that VAN DER WAALS plucked from a thistle patch of roundabout approximations by power series, empirical data, and crude geometrical arguments.

explain[3] what is meant by θ in general in that part, namely a given multiple of the volume of a body of some ideal gas at some specified pressure.

Remark 2. Such natural fluids as conform, more or less, to (2.9) do so only at sufficiently high temperatures and sufficiently low pressures. Thus in order to represent their total behavior, in some part of the positive V-θ quadrant we should have to use a relation different from (2.9). Natural gases suffer changes of phase: If cooled sufficiently, they become liquids, and the thermal equations of state of liquids are not even roughly like (2.9); if cooled still further, these liquids become solids, which do not conform to any relation of the form (2.1) except in peculiarly simple circumstances. It is not to any natural fluid that the term "ideal gas" refers. Rather, as the name suggests, an ideal gas, like a rigid body, is a concept, not a physical object.

Remark 3. That the scale of temperature selected for use in constructing the theory should be such as to give the set of all possible temperatures a finite greatest lower bound has sometimes been regarded as essential. According to SOMMERFELD,[4] "Without the fixed though inaccessible lower bound for the temperature the whole structure of thermodynamics would collapse." To set the zero of such a scale as the ideal reading corresponding to $p = 0$ for an ideal gas is merely a convention. So as to emphasize the arbitrariness of this convention, the early authors wrote (2.9) in the form $\varpi V = C(1 + \alpha\Theta)$, in which C and α are suitably selected constants, and Θ is the temperature on some scale or other, all such scales being assumed affinely related. The form (2.9) is obviously equivalent but more convenient.

3. This is the way P. S. EPSTEIN introduced temperature in §4 of Chapter 1 of his book for physicists, *Textbook of Thermodynamics*, New York, John Wiley, 1937.

The reader imbued with the folklore of thermodynamics should read *A History of the Thermometer and its Use in Meteorology* by W. E. KNOWLES MIDDLETON, Baltimore, Johns Hopkins Press, 1966, especially Chapter four, "The Search for Rational Scales". He will learn that most early experimentists presumed the existence of some absolute hotness; their problem was not to define it but to measure it correctly. He will learn also that several early savants inferred either from the measured data on real gases or from molecular notions that there must be some "absolute cold", and some of them gave figures equivalent to determining it on the empirical scales they used. MIDDLETON converts some of these to degrees C as follows:

AMONTONS (1699):	$-248°C$
LAMBERT (1779):	$-270°C$
REGNAULT (1847):	$-272.75°C$
RANKINE (1853):	$-274.6°C$.

To this we may add that CARNOT, referring to "the laws of Mariotte and Gay-Lussac", assumed in effect that absolute cold was $-267°C$. All these results were obtained by extrapolation from experiments on real gases, not from kinetic theories or metaphysical arguments about coupling bodies, *etc*. For those who wish to take "absolute" temperature as a primitive concept, those experiments provide a basis more immediate than is any available for various other primitive concepts of physics, for example, mass and force.

4. A. SOMMERFELD, § 11 of *Thermodynamik und Statistik*, herausg. F. BOPP und J. MEIXNER, Wiesbaden, 1952.

Remark 4. Those who are tempted to stir the hornets' nest encasing the "physical definition" of temperature may replace θ by $f(\theta)$, f being a sufficiently smooth, increasing function, and call $f(\theta)$ an "empirical temperature". Most, but not all, of the developments in this tractate will remain unaffected thereby.

Remark 5. Since KV will do just as well as V as a measure of temperature according to an ideal gas thermometer, K being any positive constant, $K\theta$ provides just as good a scale of temperature in terms of (2.9) as does θ. Moreover, if we assign the pressure and volume and use (2.9) to determine θ for a particular body of ideal gas, then, at the same pressure and volume, $K\theta$ is the temperature of a body of ideal gas whose constitutive constant is R/K. That is, the unit of the scale of temperature as measured by an ideal gas thermometer is arbitrary. The finite greatest lower bound, taken arbitrarily as 0, is not, for it is the ideal temperature at which no pressure at all is needed to restrain the body of ideal gas within whatever volume it may occupy. A scale of temperature of this kind is nowadays identified with an "absolute" scale, though it has nothing to do with the scale to which KELVIN first gave that name.[5]

EXAMPLE 2. The fluid body such that

$$\varpi = \frac{R\theta}{V - b} - \frac{a}{V^2}, \tag{2.10}$$

R, a, and b being positive constants, is called the body of *Van der Waals fluid* whose constitutive constants are R, a, and b. In this example $\mathcal{D}$ is a suitably

5. In republishing his paper of 1848 on this subject, KELVIN in 1881 [W. THOMSON, p. 106 of *Mathematical and Physical Papers* 1, Cambridge, 1882] wrote that while his original paper "was wholly founded on Carnot's uncorrected theory, according to which the quantity of heat taken in in the hot part of the engine...was supposed to be equal to that abstracted from the cold part...", nevertheless the required "corrections...do not in any way affect the absolute scale for thermometry which forms the subject of the present article". This claim will not bear examination unless we grant to the vague term "absolute" two entirely different meanings. According to the first, which KELVIN had used in 1848, it is possible to introduce a thermometric scale $\tau(\theta)$, τ being an increasing function, in such a way that the work done by a simple Carnot cycle is exactly $(\tau_{\max} - \tau_{\min})C^+$. This was KELVIN's original sense of "absolute". In particular, his "absolute" temperature did not bear an independent unit, for its units were those of work/heat. Moreover, it had nothing to do with "absolute zero", for the choice of the zero-point made no difference.

A temperature "absolute" in KELVIN's first sense exists according to the caloric theory. Indeed, if $\bar{g}$ is the function occurring in (10.12), below, then $\tau = \bar{g} + $ const. If, as the pioneers always assumed, $\Lambda_V > 0$ and $\partial \varpi / \partial \theta > 0$, then the work done by a simple cycle equals the area inclosed by that cycle in a plane whose co-ordinates are τ and the value of CARNOT's heat function H_1. A direct mathematical development of CARNOT's theory along the lines suggested by KELVIN's work may be found in the paper by TRUESDELL, "Theoria de effectibus mechanicis caloris iam pridem ab ill$\underline{\text{mo}}$ Sadi Carnoto verbis physicis promulgata nunc primum mathematice enucleata", *Atti dell'Accademia di Scienze dell'Istituto di Bologna, Classe di Scienze Fisiche* (12) **10** (1973), 29–41.

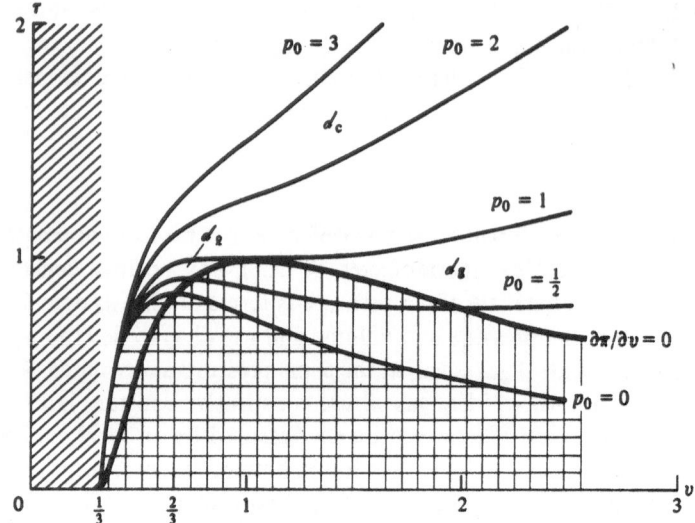

Figure 2. The regions d_o, d_g, and d_ℓ for a Van der Waals fluid.

selected proper subset of the quadrant defined by the inequalities $V > b$, $\theta > 0$. The equation itself is regarded as a model for both the gaseous and the liquid phases of a real fluid body. Like a body of ideal gas, a body of Van der Waals fluid has no piezotropic points. Its character is revealed economically by use of reduced, dimensionless variables:

$$\pi \equiv \frac{\varpi}{a/(27b^2)}, \qquad v \equiv \frac{V}{3b}, \qquad \tau \equiv \frac{\theta}{(8a)/(27Rb)}. \tag{2.11}$$

Then (2.10) assumes the form

$$\pi = \frac{8\tau}{3v - 1} - \frac{3}{v^2}. \tag{2.12}$$

Figure 2 presents some of the features of this equation. The diagonally hatched region of the v-τ quadrant, in which $v \leqq \frac{1}{3}$, is excluded by definition; the domain d of (2.12) is some proper subset of the remainder of the quadrant. The vertically hatched region bounded above by the heavy line which is the locus of points at which $\partial\pi/\partial v = 0$, is the region in which (2.2) is violated. It is usually dismissed as being "unstable" and "without physical reality". Elsewhere, (2.4) shows that τ is an increasing function of v on an isobar. In further description of Figure 2, though not elsewhere, we denote by the term "isobar" the entire locus $\pi(v, \tau) = p_0 = $ const. Likewise, we shall call the lines $\tau = $ const. "isotherms". The domain above the isobar for which $p_0 = 1$ is denoted by d_o. The isobars for which $p_0 > 1$ intersect each isotherm at exactly one point. On such an isobar, which is maximal, the temperature determines one and only one volume. Points in the domain d_o correspond to the constitutive domain of a common phase which may be regarded at pleasure as being either gaseous or liquid. If $0 < p_0 < 1$, however, each isobar is composed of two disconnected maximal isobars lying in the region where $\partial\pi/\partial v < 0$, and some isotherms intersect both of these. That is, the entire isobar intersects some isotherms at two distinct volumes, of which the larger is regarded as appropriate to the gaseous phase, the smaller to the liquid. Likewise, the isotherms on which $\tau < 1$ have two disconnected parts in the domain where (2.2) is satisfied. The regions corresponding to the two phases in the quadrant of the reduced variables v and τ are indicated by d_g and d_ℓ, respectively. The region d_ℓ is bounded; indeed, as is consonant with the behavior of liquids, within it the volume is allowed to vary relatively little: $\frac{1}{3} < v < 1$. If we set aside the requirement that $\partial\pi/\partial v < 0$, then along the isobars for which $p_0 < 1$ the temperature experiences extremes at the points where $\partial\pi/\partial v = 0$.

Values of θ, V, and p that correspond to values of τ, v, and π that are less than 1 are called *subcritical*.

The horizontally hatched region below the isobar for which $p_0 = 0$ corresponds to negative pressures. The part of this that lies outside the vertically hatched region is sometimes set aside, sometimes regarded as representing a liquid supporting tension instead of pressure. If we exclude this region from d_ℓ, then the inequalities (2.2) and (2.5) are satisfied in d_ℓ as well as in d_g and d_c.

Henceforth we shall regard the region above the line $\partial\pi/\partial v = 0$ as representing the constitutive domain of a body of Van der Waals fluid, for in this domain (2.2) is satisfied. For this body the fact that some isobars and isotherms have two disconnected components in the constitutive domain must be noticed but will not occasion us any trouble.

Scholion. As we see by a glance at Figure 2, (2.1) cannot always be inverted for V. Thus this fluid body serves as an example to show that in thermodynamics we must not confuse local invertibility with invertibility in the large. It shows also why *we cannot use p and θ as independent variables* in constructing a comprehensive thermodynamics, however convenient these variables may be in applications to many particular bodies.

(The interpretation of the Van der Waals equation in the context of changes of phase is familiar and need not be repeated here.)

EXAMPLE 3 (BRIDGMAN's *"Hypothetical Liquid Water"*). Among fluids having piezotropes, water seems to be the only one for which there are many data. Figure 3 reproduces a graph published by BRIDGMAN[6] so as to

6. P. W. BRIDGMAN, "Water, in the liquid and five solid forms, under pressure", *Proceedings of the American Academy of Arts and Sciences* 47 (1911/12), 441–558 (1912) = *Papers* 1, 213–333. In the range of temperatures considered here the revised data BRIDGMAN published shortly afterward differ little from those he used so as to compose the graph reproduced above in Figure 3. These later data are given in his "Thermodynamic properties of liquid water to 80° and 12000 kgm", *ibid.* 48 (1912/13), 309–362 (1912) = *Papers* 1, 379–432.

The data at temperatures far below those ordinarily regarded as freezing points at the pressures indicated, BRIDGMAN explained as follows: ". . .it is the easiest possible matter

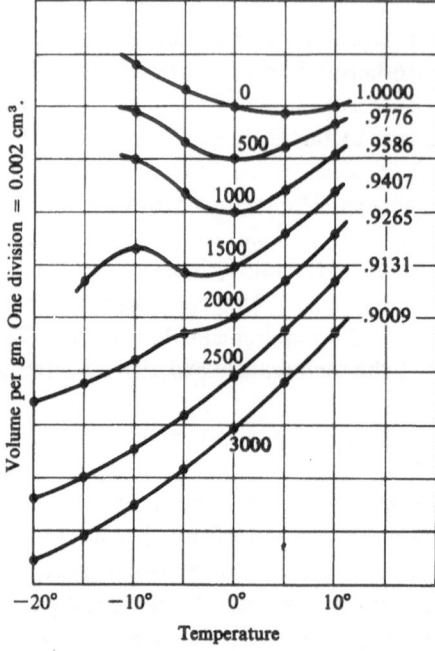

Figure 40. Curves showing the relation between volume and temperature of water for various constant pressures. The numbers in the body of the diagram show the constant pressure of each curve; the numbers to the right show the volume at 0° of the liquid at the indicated pressure. If drawn to scale the curves should be separated about ten times as much as shown.

Figure 3. Isobars of water at low temperatures according to BRIDGMAN.

display the behavior of water at low temperatures and high pressures. The reader must note his statement that the curves, though each one is separately drawn to the scale indicated, would have to be spaced about ten times further apart, were the graph itself to be put into scale. BRIDGMAN was particularly pleased with the results at 1500 atmospheres overpressure in that they displayed the S-like form he had sketched (Figure 39, p. 543) as being the "hypothetical relation . . . for liquid water if it could be subcooled indefinitely without freezing".

On a scale drawing in the V-θ quadrant the minima of volume on the

---

to subcool the liquid with respect to any one of the four solid phases bordering on the domain of the liquid". See p. 531 of his earlier paper.

The minima of volume on isobars for pressures between 1 atm and 200 atm had been determined earlier by AMAGAT. His results are shown in Figure 5, p. 18, of the book by G. TAMMANN, *Über die Beziehungen zwischen den inneren Kräften und Eigenschaften der Lösungen*, Hamburg und Leipzig, Voss, 1907. Measurements, all over half a century old, of the temperature of maximum density at pressures from −26.3 atm to 600 atm, are collected in Table 133 on p. 277 of the book by N. E. DORSEY, *Properties of Ordinary Water-Substance*, New York, Reinhold, 1940.

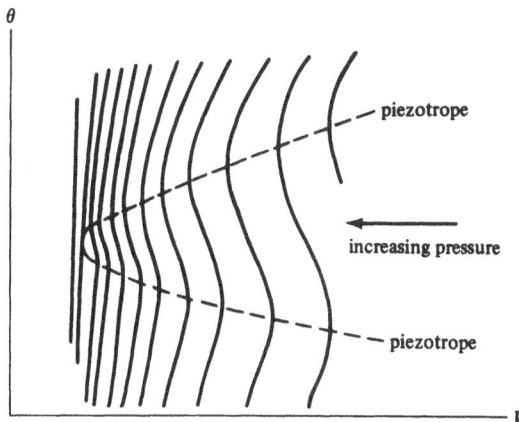

Figure 4. Possible field of isobars of BRIDGMAN's "Hypothetical . . . Liquid Water".

isobars at overpressures 0, 500, 1000, and 1500 seem to lie on a curve which descends as the volume decreases and ends at one of the two points where the isobar at 2000 atmospheres as sketched shows an inflection. Curves like that of BRIDGMAN's "hypothetical relation", arranged as his data suggest they may be, are shown in Figure 4. The result following (2.8) shows that $\partial^2\varpi/\partial\theta^2 \geqq 0$ on the upper branch of the piezotrope, $\partial^2\varpi/\partial\theta^2 \leqq 0$ on the lower branch, and $\partial^2\varpi/\partial\theta^2 = 0$ at the point where the two join. Moreover, the figure as drawn makes the temperature on the upper branch a decreasing function of pressure. If $\partial^2\varpi/\partial\theta^2 \neq 0$ on the upper branch, the fourth equation in Footnote 2 requires that $\partial^2\varpi/\partial V\partial\theta \leqq 0$ on that branch. It is easy to verify that $\partial^2\varpi/\partial V\partial\theta \leqq 0$ on the lower branch, too, again on the assumption that $\partial^2\varpi/\partial\theta^2$ does not vanish on it.

The array, like BRIDGMAN's single curve, is hypothetical and is intended only as an illustration of the possibilities thermodynamics must be broad enough to allow. If the diagram is supposed to extend arbitrarily far into the region of low pressures, the "hypothetical water" it represents could be superheated indefinitely without boiling. That real water has no piezotropic points at pressures much below one atmosphere, is easily allowed for by supposing $\mathscr{D}$ to be so limited that the isobars of the liquid phase stop short before reaching the upper branch of the piezotrope from above.

In later experiments BRIDGMAN[7] failed to find the S-shaped curve even once, or indeed any piezotropic points at all. Still later BRIDGMAN[8] found a systematic error in all of his earlier measurements of pressures. Whether this error

7. P. W. BRIDGMAN, "The pressure-volume-temperature relations of the liquid, and the phase diagram of heavy water", *Journal of Chemical Physics* 3 (1935), 597–605 = *Papers* 5, 2919–2927. There BRIDGMAN wrote,

With regard to the values below 0°, this is the first time since 1912 that the volume has been measured in this region. The results now found do not check in fine detail

had anything to do with the discrepancies between his two sets of data for water, he did not state.

For the theorist who would found his theory on details revealed by experiment, the watery data provided over a period of decades by this Nobel laureate and advocate of Operationalism should provide a classic warning that he may build upon mud.

R.J. JAWORSKI & J.H. KEENAN have provided a thorough review and analysis of the properties of water near the piezotrope, which they find to be a straight line: ''Thermodynamic properties of water in the region of maximum density'', *Journal of Applied Mechanics* **89** (1967), 478–483. By adjustment to data, especially AMAGAT's, which are given in DORSEY's book (cited above on p. 16), and by use of a method devised earlier by KEENAN, they obtain a complicated equation of state expressed in particular units.

All data we have seen are compatible qualitatively with Figure 4 *if the boundary of 𝒟 be properly set*: crossing the upper branch of the piezotrope from above at a pressure just a little below 1 atm, and for high pressures remaining just a little below that branch.[9]

Scholion 1. BRIDGMAN's "Hypothetical . . . Liquid Water" shows that *we cannot use p and V as independent variables* if we are to construct a comprehensive thermodynamics. The famous axiomatization of thermodynamics by CARATHÉODORY, endorsed by BORN and other physicists, begins by supposing the contrary.[10] So do many textbooks.[11]

Scholion 2. For definiteness we always interpret V as the volume of a fluid body and p as the pressure required to maintain that body at that volume when the temperature is assigned. As far as the formal theory goes, however, p and V might just as well stand for other physical quantities. For example, V might be the length of a filament and p the pressure or tension required to

---

with those found before; in particular the minimum and maximum of volume as a function of temperature on the isobar at 1500 kg found in 1912 and shown in Fig. 40 of the 1912 paper, has [sic] not been found this time. Irregularities are now found suggesting the maximum and minimum, but not so pronounced.

8. P. W. BRIDGMAN, pp. 416–417 of the supplement subjoined to *The Physics of High Pressure*, London, G. Bell & Sons, 1949, and later editions.

9. One of the few discussions of the matter to be found in recent books is that of J. KESTIN, §§ 3.3, 7.4.9, and 7.4.10 of his *A Course in Thermodynamics*, Waltham, Blaisdell, 1966. It is largely schematic. The curves drawn are like the upper parts of those in Figure 4 but more angular, with the portions below the upper part of the piezotrope short and nearly parallel to the V-axis. The range of pressures is 0.008 atm to 270 atm.

10. *Cf.* the concluding remarks of L. A. TURNER, "Further remarks on the zeroth law", *American Journal of Physics* **30** (1962), 804–806; J. S. THOMSEN & T. J. HARTKA, "Strange Carnot cycles; Thermodynamics of a system with a density extremum", *ibid.* 26–33, 388–389.

11. See, for example, §2.5 of the book by KESTIN, cited in Footnote 9.

maintain that length when θ is given a fixed value. Of course, various inequalities may be reversed or may fail altogether for different physical
interpretations of the letters V and p.

Theorem 1 (First Reversal Theorem).   *Let a path $\mathscr{P}$ be given. Then a
given body does the same work $L(\mathscr{P})$ in all processes of $\mathscr{P}$. Also*

$$L(\mathscr{P}_1 + \mathscr{P}_2) = L(\mathscr{P}_1) + L(\mathscr{P}_2), \tag{2.13}$$

$$L(-\mathscr{P}) = -L(\mathscr{P}). \tag{2.14}$$

*If $\mathscr{C}$ is a simple cycle oriented clockwise, and if the region $\mathscr{A}$ inclosed by it is
contained in $\mathscr{D}$, then*

$$L(\mathscr{C}) = \iint_{\mathscr{A}} \frac{\partial \varpi}{\partial \theta} \, dV d\theta. \tag{2.15}$$

Proof.   Axiom I implies that in a given process the pressure upon a
given body is determined as a function of time: $p(t) = \varpi(V(t), \theta(t))$.
All the results except (2.15) follow from (1.6) and Definition 8 by use
of elementary properties of integrals. Then (2.15) follows by
AMPÈRE's transformation of an integral around a simple plane
circuit into an integral over the region inclosed by it. $\square$

Historical Comment.   AMPÈRE's transformation seems not to have become
widely known until some time after he published it (1826). The special result
(2.15) in differential form was derived roughly by CLAPEYRON (1834).

CHAPTER 3

The Doctrine of Latent and Specific Heats. Second Reversal Theorem. Fundamental Theorem of Calorimetry.

Axiom II (Doctrine of Latent and Specific Heats). *The heating Q of a given body in any process V, θ is determined as follows by means of functions Λ_V and K_V, defined over $\mathscr{D}$, continuous and having continuous partial derivatives there:*

$$Q = \Lambda_V(V, \theta)\dot{V} + K_V(V, \theta)\dot{\theta} \tag{3.1}$$

at all times when $\dot{V}$ and $\dot{\theta}$ exist. Moreover,

$$K_V > 0. \tag{3.2}$$

Historical Comment. Studies of the history of thermodynamics have failed, generally, to notice that this axiom was accepted, if often only tacitly, by every early author of thermodynamic research. The earliest fairly clear statements of it seem to be those of CLAUSIUS (1850) and KELVIN (1851). The ideas it represents were distilled from the experiments on calorimetry in the late eighteenth century. See the book by D. MCKIE & N. DE V. HEATHCOTE, *The Discovery of Specific and Latent Heats*, London, Arnold, 1935.

Remark 1. The functions Λ_V and K_V are called the *latent heat with respect to volume* and the *specific heat at constant volume*, respectively. They are the third and fourth of the four constitutive quantities in the substructure upon which classical thermodynamics rests, the other two being $\mathscr{D}$ and ϖ.

The term "latent heat" has survived in the modern literature only in connection with fusion and evaporation, or their contraries, solidification and condensation. The latent heat Λ_V that occurs in (3.1) does not refer to melting or boiling, and all results we shall obtain below are intended to apply only in a range of temperatures and volumes (or pressures) such as to exclude changes of phase.

We use the term "specific heat" in the sense introduced by BLACK. Later LAVOISIER and LAPLACE employed "specific heat" and "heat capacity" interchangeably, referring both to unit mass. Nowadays, usually only the former term is so referred; the "heat capacity" is what we here denote by K_V, and the "specific heat" of a body of mass M is K_V/M. We treat always a single body of fixed mass, and as units are arbitrary, we do no violence to the concepts in reverting to the earliest verbal usage, which several of the pioneers of thermodynamics maintained.

Remark 2. The constitutive inequality (3.2) asserts that *in order to raise the temperature of a body without changing its volume, we must supply heat to it.*

Remark 3. Because of Definition 2 in Chapter 1, Q is defined at t_1, at t_2, and at all but a finite number of times in $]t_1, t_2[$. Further, it is integrable on $[t_1, t_2]$.

Theorem 2 (Second Reversal Theorem). *Let a path $\mathscr{P}$ be given. The net gain of heat by a given body is the same in all the processes of $\mathscr{P}$. Let $C(\mathscr{P})$ denote this net gain of heat. Then*

$$C(\mathscr{P}_1 + \mathscr{P}_2) = C(\mathscr{P}_1) + C(\mathscr{P}_2), \tag{3.3}$$

$$C(-\mathscr{P}) = -C(\mathscr{P}). \tag{3.4}$$

In every process of a path $\mathscr{P}$, a given body absorbs the same amount of heat $C^+(\mathscr{P})$ and emits the same amount of heat $C^-(\mathscr{P})$. Further,

$$C^+(-\mathscr{P}) = C^-(\mathscr{P}), \qquad C^-(-\mathscr{P}) = C^+(\mathscr{P}). \tag{3.5}$$

If $\mathscr{C}$ is a simple cycle oriented clockwise, and if the region $\mathscr{A}$ it incloses is contained in $\mathscr{D}$, then

$$C(\mathscr{C}) = \iint\limits_{\mathscr{A}} \left(\frac{\partial \Lambda_V}{\partial \theta} - \frac{\partial K_V}{\partial V} \right) dV d\theta. \tag{3.6}$$

Proof. The theorem is an immediate consequence of Axiom II, and the proof parallels that of Theorem 1. □

Remark 1. The two reversal theorems suggest that in order to exploit the consequences of Axioms I and II it should suffice to consider simple paths only. Any parts of a path that are reverses of other parts may be discarded as far as calculations of heat and work are concerned.

Remark 2. Although CLAUSIUS gave a painful and imprecise proof of (3.6) in 1850, it results straight off by an application of AMPÈRE's transformation. Since Axiom II makes Λ_V and K_V have continuous derivatives in $\mathscr{D}$, use of AMPÈRE's transformation is justified.

Historical Comment. Although we can find no clear early statement of either theorem of reversibility, both were well known in principle to all the early authors. CARNOT, for example, wrote[1] in connection with his cycle

> The operations I have just described could have been effected in an inverse sense and inverse order.

General Scholion. One of the senses in which the term "reversible process" is used in works on thermodynamics means neither more nor less than the assertions (2.14) and (3.4) of the two reversal theorems. These theorems follow from Axioms I and II, irrespective of the choice of the four constitutive quantities $\mathscr{D}$, ϖ, Λ_V, and K_V. Although those axioms were abundantly general for the purposes of classical thermodynamics, in modern thermomechanics they express severe restrictions upon the class of bodies allowed to the theory. We must regard Axioms I and II themselves as defining a certain *constitutive class*, which, while providing the framework upon which classical thermodynamics is constructed, is a highly special one if seen in the light of modern thermomechanics. It is misleading to speak of processes as being "reversible" or "irreversible" in themselves, for it is the constitutive relations of a body that determine whether, for that body, a process be reversible or not reversible. The constitutive class presumed in classical thermodynamics is such as to restrict attention to *bodies for which all processes are reversible.*

In order to consider "irreversible processes" at all—except in the phantasmagoria for which traditional thermodynamics is famous—it is necessary to broaden the constitutive class laid down at the outset in the classical treatments. No such thing will be done here. Our only objective is to reveal the classical theory as rational and selfcontained, free of metaphysics and divination.

Corollary 2.1. *On a simple path $\mathscr{P}_\theta$ consisting of isothermal processes with initial point (V_a, θ) and final point (V_b, θ), if $(V_b - V_a)\Lambda_V(V, \theta) \geqq 0$ on the isothermal segment joining the initial point with the final point, then*

$$C^+(\mathscr{P}_\theta) = \int_{V_a}^{V_b} \Lambda_V(V, \theta)dV. \qquad (3.7)$$

In particular, heat is absorbed in an isothermal expansion in a part of $\mathscr{D}$ where $\Lambda_V > 0$ and emitted in a part where $\Lambda_V < 0$.

Corollary 2.2. *In a part of $\mathscr{D}$ where Λ_V is of one sign, if $C^+(\mathscr{P}_\theta) \to 0$ on a sequence of simple paths $\mathscr{P}_\theta$ consisting of isothermal processes with fixed initial point (V_a, θ) such that $\operatorname{sign}(V_b - V_a) = \operatorname{sign} \Lambda_V$, then $V_b - V_a \to 0$.*

---

1. *Réflexions*, p. 19. See also p. 35, and the argument about maximum efficiency on pp. 21–22 and 35–36:

> The result of the first operations was the production of a certain quantity of motive power and the transport of some of the caloric of body A to body B; the result of the inverse operations is the consumption of the motive power produced, and the return of the caloric from body B to body A, in such a way that the two sequences of operations annul each other, in some way neutralize each other.

Proof. Because Λ_V is of one sign, $C^+(\mathscr{P}_\theta)$ for a fixed value of V_a is an increasing or decreasing function of V_b. □

Historical Comment. The pioneers of thermodynamics always assumed that $\Lambda_V > 0$. CARNOT, who limited his considerations to gases, all of which he considered to be ideal or very nearly so, wrote (*Réflexions*, pp. 29–31)

> When a gaseous fluid is rapidly compressed, its temperature rises; on the contrary, its temperature falls when it is rapidly expanded. This is one of the facts best confirmed by experiment. We shall take it as the basis of our proof.

Those sentences mean that if $Q = 0$ and $\dot{V} \neq 0$, then sign $\dot{\theta} = -\text{sign } \dot{V}$. That is, by (3.1), $\Lambda_V/K_V > 0$. In the following sentences of the same passage CARNOT states in effect that $K_V > 0$.

For some substances, water in the range of its "anomalous behavior" being one, the opposite seems to hold: $\Lambda_V < 0$ for some values of V and θ.

Theorem 3 (Fundamental Theorem of Calorimetry). *Almost always*[2]

$$Q = \Lambda_p \dot{p} + K_p \dot{\theta}; \tag{3.8}$$

Λ_p, *the latent heat with respect to pressure, and* K_p, *the specific heat at constant pressure, are related as follows to* Λ_V *and* K_V:

$$\Lambda_p = \frac{\Lambda_V}{\dfrac{\partial \varpi}{\partial V}}, \qquad K_p - K_V = -\Lambda_V \frac{\dfrac{\partial \varpi}{\partial \theta}}{\dfrac{\partial \varpi}{\partial V}}. \tag{3.9}$$

Also

$$(K_p - K_V)Q = -\Lambda_V \left[\frac{K_V}{\dfrac{\partial \varpi}{\partial V}} \dot{p} - K_p \dot{V} \right] \tag{3.10}$$

almost always.

Proof. From (2.1) we see that almost always

$$\dot{p} = \frac{\partial \varpi}{\partial V} \dot{V} + \frac{\partial \varpi}{\partial \theta} \dot{\theta}. \tag{3.11}$$

Because of (2.2), we may eliminate $\dot{V}$ between this relation and (3.1) so as to get (3.8) and (3.9). To obtain (3.10), we calculate $K_p K_V \dot{\theta}$ from (3.8) and (3.1), subtract the results, and use the first equation of (3.9). □

Remark. It is commonly assumed that ϖ is invertible globally, not only locally as implied by (2.2), to yield V as a function of p and θ. In that case K_p and Λ_p in (3.8) are equal to functions of p and θ. In this tractate we shall not

---

2. Henceforth we shall use the term "almost always" to mean "except, possibly, at a finite number of times".

need to assume such invertibility except for some of the considerations in Chapter 14.

Corollary 3.1. *The latent heats Λ_p and Λ_V are always of opposite sign.*

Proof. The assertion follows at once from the first equation of (3.9) and from (2.2). □

Historical Comment. CARNOT (*Réflexions*, p. 31) assumed that $\Lambda_p < 0$. Whether or not he knew the relation expressed by the first equation of (3.9), which makes this statement follow from his earlier assumption that $\Lambda_V > 0$, is not clear.

Ideal gases as defined by (2.9) will serve us often as examples, and then in Part III we shall use them as essential elements in building the final theory. Two of their simplest properties may be read off as corollaries of (3.9).

Property 1. For a body of ideal gas

$$K_p - K_V = \frac{R\Lambda_V}{\varpi}. \tag{3.12}$$

Property 2. If the difference of the specific heats of an ideal gas is a positive constant, there is a positive constant J such that

$$J\Lambda_V = \varpi = \frac{R\theta}{V}. \tag{3.13}$$

Conversely, if (3.13) holds for some positive constant J, then

$$J(K_p - K_V) = R. \tag{3.14}$$

Historical Scholion. The condition (3.14) was inferred by J. R. MAYER in 1842 on partially metaphysical grounds. The condition (3.13) was put forward by HOLTZMANN in 1845 as being plausible on the basis of then known or conjectured properties of gases. It asserts that to within a factor J that converts caloric units into mechanical ones, the latent heat of a body of ideal gas is just the pressure acting on that body. That the Doctrine of Latent and Specific Heats makes MAYER's Assumption and HOLTZMANN's equivalent, seems neither to have been known to those pioneers and their immediate successors nor to be commonly known today.

It seems to be believed widely that MAYER's Assumption (3.14) reflects the uniform and universal interconvertibility of heat and work or the "First Law of Thermodynamics". This belief is untrue. In the first Historical Scholion of Chapter 11 we shall demonstrate its falsity and draw consequences for the history of thermodynamics.

Remark. Let γ denote the *ratio of specific heats:*

$$\gamma \equiv \frac{K_p}{K_V}. \tag{3.15}$$

It is a consequence of the second equation of (3.9) that $\gamma = 1$ at a piezotropic point or a point where $\Lambda_V = 0$. For a body of ideal gas we infer from (3.12) that

$$\frac{V\Lambda_V}{K_V} = (\gamma - 1)\theta. \tag{3.16}$$

CHAPTER 4

Adiabatic Processes.
Laplace's Theorem.

Definition 11. A process is *adiabatic* if $Q(t) = 0$ at all times t at which $Q(t)$ is defined. Any curve in $\mathscr{D}$ whose tangent vector in some parametric representation is always orthogonal to the vector field (Λ_V, K_V) is called an *adiabat*.

Remark 1. From the definition, it is clear that a process V, θ of a given body is adiabatic if and only if it satisfies

$$\Lambda_V \dot{V} + K_V \dot{\theta} = 0 \tag{4.1}$$

almost always. Thus the concept of an adiabatic process is a constitutive one. A process that is adiabatic for one body is not so, generally, for another. Likewise, the adiabats of one body generally differ from those of another. A constant process is adiabatic as well as isothermal and isochoric.

Remark 2 (Existence and Uniqueness of Adiabats). By definition, a parametric representation of an adiabat satisfies

$$\Lambda_V V' + K_V \theta' = 0. \tag{4.2}$$

Because $K_V > 0$, by reasoning parallel to that after (2.3) we conclude that along an adiabat θ is a twice continuously differentiable function of V satisfying

$$\frac{d\theta}{dV} = -\frac{\Lambda_V}{K_V}. \tag{4.3}$$

Conversely, let Λ_V and K_V be the latent and specific heats of a certain body, and consider any solution of (4.3) on an interval of volumes. The curve represented by this solution is easily seen to be an adiabat for the body. Thus (4.3) is a differential equation for determining the adiabats of a body. Because Λ_V and K_V have continuous partial derivatives in $\mathscr{D}$, CAUCHY's Existence Theorem[1] assures us that through every point of $\mathscr{D}$ runs one and only one adiabat, uniqueness being understood in this sense: If two adiabats passing through a point are defined on the same interval of volumes, they are the same. The *maximal adiabat* through a given point is defined in the same way as the maximal isobar in Remark 1 after Axiom I in Chapter 2.

Using (4.1), we see that in a simple adiabatic process θ is a twice continuously differentiable function of V which satisfies the differential equation (4.3) of an adiabat. Thus a simple adiabatic process traverses an adiabat. Further, from any solution of (4.3) on a bounded, closed interval of volumes we can easily construct adiabatic processes which traverse the adiabat defined by the solution.

Because $d\theta/dV$ is finite, *an adiabat is never orthogonal to an isotherm*. Moreover, because Λ_V and K_V have continuous partial derivatives, the curvature field of the adiabats exists and is continuous.

Definition 12. A point of $\mathscr{D}$ where $\Lambda_V \neq 0$ is an *ordinary point*. A point where $\Lambda_V = 0$ is a *neutral point*. A curve of neutral points is a *neutral curve*; a nonempty open set of neutral points is a *neutral part* of $\mathscr{D}$.

Remark 1. At a neutral point $\Lambda_p = 0$, and $\gamma = 1$, as we see at once from (3.9) and (3.15).

Remark 2 (Decussation of Adiabats and Isotherms near an Ordinary Point). Because Λ_V is continuous, each ordinary point of $\mathscr{D}$ has a neighborhood consisting of points at which Λ_V is of one sign. In every such neighborhood, if sufficiently small, each point is the intersection of one adiabat with one isotherm. As we shall say, *every ordinary point has a neighborhood in which the adiabats and isotherms decussate*. This simple fact is of central importance in many of the later arguments. We might, if we liked, in such a

---

1. A.-L. CAUCHY, "Mémoire sur l'intégration des équations différentielles", pp. 327–384 of *Exercices d'Analyse et de Physique Mathématique* 1 (1840) = *Oeuvres* (II) 11, 399–465, being a reprint of a lithographed publication of 1835. CAUCHY presented this theorem to his classes at the École Polytechnique, 1820–1830. The proof is based on a scheme of successive approximation introduced by EULER in his famous treatise on integral calculus (1768).

The theorem of CAUCHY is a local result. On the contrary, in all the early writings on thermodynamics we have found ubiquitous childlike faith in the algebra of calculus, with no sliver of a shadow of apprehension lest the result of any juggling of *d*s and ∫s might fail to yield a global truth of physics.

A modern, stronger version of CAUCHY's theorem is Theorem 2.3 of Chapter I of E. A. CODDINGTON & N. LEVINSON, *Theory of Ordinary Differential Equations*, New York *etc.*, McGraw-Hill, 1955.

neighborhood use adiabats and isotherms as co-ordinate curves, but we shall not do so.

Historical Comment. The early writers seem to have presumed that isothermal segments and adiabats could serve as co-ordinate curves through-out $\mathcal{D}$. REECH expressly assumed such was the case, and his drawings show that he always envisaged the adiabats as being like those of an ideal gas (Figure 5, below). In each neighborhood of a point on a neutral curve separating a region in which $\Lambda_V > 0$ from one in which $\Lambda_V < 0$ the adiabats generally intersect some isotherms at least twice. Example B of Figure 7 illustrates this possibility. Therefore, *the classical tacit assumption is generally false.*[2] Fortunately that assumption is not necessary to the general theory of thermodynamics, as we shall see.

Remark 3 (Neutral Points, Neutral Curves). If, at a time when Q is defined, a process occupies a neutral point, then (3.1) reduces to $Q = K_V \theta$. At such a time, by (3.2), the sign of Q is the sign of θ; that is, if the heating is positive the temperature is increasing, and conversely. We may therefore say in particular that in a process that traverses a neutral curve, heat is always absorbed as the temperature increases, emitted as the temperature decreases. As a glance at (4.3) shows, *at a neutral point, the adiabat cuts the isochor orthogonally and is tangent to the isotherm. At all other points on an adiabat,* $d\theta/dV \neq 0$. Therefore, *along an adiabat, the temperature passes through extremes at neutral points and at them only.* Thus a neutral curve is the locus of extremes of temperature along adiabats. A neutral curve is an adiabat if and only if it is also an isothermal segment.

By differentiating (4.3) we easily find the curvature of the adiabat at a neutral point:

$$\frac{d^2\theta}{dV^2} = -\frac{1}{K_V}\frac{\partial \Lambda_V}{\partial V}. \tag{4.4}$$

Theorem 4 (LAPLACE, 1816–1823, in principle). *Along an adiabat p is a continuously differentiable function of V, and*

$$\frac{dp}{dV} = \gamma \frac{\partial \varpi}{\partial V}, \tag{4.5}$$

γ *being defined by* (3.15).

Proof. Remark 2 after Definition 11 shows that on an adiabat θ is a continuously differentiable function of V. Hence by (2.1) p is a function of V on an adiabat. By (4.3) this function satisfies

$$\frac{dp}{dV} = \frac{\partial \varpi}{\partial V}\left(1 - \frac{\Lambda_V}{K_V}\cdot\frac{\partial \varpi/\partial \theta}{\partial \varpi/\partial V}\right), \tag{4.6}$$

and so (4.5) follows by use of the second equation of (3.9). □

2. The first author to remark this fact and to see that it renders some of the classical demonstrations lacunary seems to have been POINCARÉ; see §§150–153 of his *Thermodynamique* (1888/9), Paris, Carré, where he makes a pass at rectifying one of them.

Corollary 4.1.   *In a part of $\mathcal{D}$ where γ is of one sign, θ is a continuously differentiable function of p along each adiabat, and*

$$\frac{d\theta}{dp} = -\frac{\Lambda_p}{K_p}. \tag{4.7}$$

Proof.   The result follows from (4.5), (4.3), and the first equation of (3.9). $\square$

Remark 1.   Clearly the relations (4.5) and (4.7) are satisfied in any process that traverses an adiabat. Such a process is nonconstant and adiabatic.

Remark 2.   Let ρ stand for the density of a body of fluid having positive mass M: $\rho \equiv M/V$. A hydrodynamical theorem of EULER asserts that in any region of flow such that $p = f(\rho)$, the squared speed of sound is $dp/d\rho$. The result (4.5) shows that if the sonic motion conforms to a process which traverses an adiabat and if $\varpi(\rho, \theta) \equiv \varpi(M/\rho, \theta)$, then the squared speed of sound is $\gamma \partial \varpi(\rho, \theta)/\partial\rho$. Therefore, γ being assumed positive, for an ideal gas the speed of adiabatic sound is $\sqrt{\gamma R\theta/M}$. Hence *the speed of adiabatic sound in a body of ideal gas is proportional to $\sqrt{\theta}$ if and only if $\gamma = $ const.*

Remark 3.   No relation between heat and work has yet been introduced. Thus the truth of Theorem 4 is in no way restricted by any such relation as may be imposed later. It is a theorem of calorimetry, not of thermodynamics. For this reason, in the previous remark we have had to assume γ positive rather than prove it to be so.

Corollary 4.2.   *A process undergone by a body of ideal gas for which $\gamma - 1$ is a nonvanishing function of temperature alone is adiabatic if and only if*

$$V \exp \int_{\theta_0}^{\theta} \frac{du}{u[\gamma(u) - 1]} = \text{const.,} \tag{4.8}$$

θ_0 being a fixed temperature that corresponds to some point of $\mathcal{D}$.

Proof.   Using (3.16) in (4.1), we see that a process of an ideal gas is adiabatic if and only if

$$(\gamma - 1)\frac{\dot{V}}{V} + \frac{\dot{\theta}}{\theta} = 0 \tag{4.9}$$

almost always. From (4.9), (4.8) immediately follows. $\square$

Corollary 4.3 (LAPLACE and POISSON, in principle).   *A process undergone by a body of ideal gas for which $\gamma = $ const. is adiabatic if and only if*

$$\theta V^{\gamma-1} = \text{const.;} \tag{4.10}$$

equivalently, if and only if

$$pV^{\gamma} = \text{const.} \tag{4.11}$$

Proof. Integration of (4.9) leads to (4.10). Using (2.9) in (4.10), we get (4.11). □

Remark 1. Corollary 4.3 is included as a part of a more general result: *If one of the following conditions is satisfied for a process undergone by a body of ideal gas, then the remaining two become equivalent:*

1. $\gamma = $ const. (*during the process*).
2. $\theta V^{\gamma-1} = $ const.
3. *The process is adiabatic.*

The proof is left to the reader. This statement makes it clear that Corollary 4.3 remains valid if in its statement the condition that $\gamma = $ const. for the body is replaced by the weaker condition that $\gamma = $ const. during the process.

Remark 2. The following conclusions, similar to Corollaries 4.2 and 4.3, are easy consequences of (3.16) and (4.3). For a body of ideal gas such that $\gamma - 1$ is a nonvanishing function of temperature alone, a curve in $\mathscr{D}$ is an adiabat if and only if (4.8) holds on it. For a body of ideal gas such that $\gamma = $ const., a curve in $\mathscr{D}$ is an adiabat if and only if (4.10) or its equivalent (4.11) holds on it.

Remark 3. Since the adiabats of any body satisfy (4.3), in order to determine them we must specify Λ_V as well as K_V. At first the last conclusion in Remark 2 seems not to refer to Λ_V, but that is an illusion, for (3.16) shows that if $\gamma = $ const., then $V\Lambda_V/(\theta K_V) = $ const.

Furthermore, suppose that the adiabats of a body, not necessarily a body of ideal gas, be the curves $\theta V^\alpha = $ const. for some constant α. Comparison with (4.3) gives the necessary and sufficient condition

$$\alpha\theta K_V = V\Lambda_V. \tag{4.12}$$

Comparison with (3.16) then shows that (4.12) is satisfied by a body of ideal gas if and only if $\alpha = \gamma - 1$.

We can also prove a result similar to that stated in Remark 1: *If one of the following conditions is satisfied for a curve in the constitutive domain of a body of ideal gas, then the remaining two become equivalent:*

1. $\gamma = $ const. (*along the curve*).
2. $\theta V^{\gamma-1} = $ const.
3. *The curve is an adiabat.*

From the foregoing statement we might expect that the truth of (4.10) or (4.11) on every adiabat of a body of ideal gas would imply that $\gamma = $ const., for that body. Such is the case according to the thermodynamics of CLAUSIUS, as we shall see in Remark 5 after Theorem 15 in Chapter 15, but the Doctrine of Latent and Specific Heats by itself is far from sufficient to exclude the possibility that γ shall have different constant values on different adiabats. An example suffices to make the point: With J being any positive constant and R being the constitutive constant of the ideal gas in question,

$$\Lambda_V = \frac{R\theta}{JV},$$

$$K_V = (D + C \log V)\frac{\theta}{h}, \tag{4.13}$$

$$h = (-\frac{CJ}{R} \log \theta + E)\theta.$$

The constants C, D, E, and the domain $\mathscr{D}$ are so chosen as to ensure that $K_V > 0$ and $h > 0$. The relation expressed by the first equation of (4.13) is HOLTZMANN's Assumption (3.13). By Property 2 in Chapter 3 we see that in this example MAYER's Assumption (3.14) is satisfied. Thus, for this example, the *difference* of specific heats is constant, the LAPLACE-POISSON Law (4.10) holds for adiabatic processes as well as adiabats, but γ is not constant unless $C = 0$.

More generally, we may ask for conditions such as to make the LAPLACE-POISSON Law (4.10) for adiabats compatible with MAYER's Assumption (3.14). The result stated after (4.12) shows that γ, and hence K_V, must then be constant along every adiabat. Use of (4.3) and (3.13) leads to the following partial differential equation for K_V:

$$JVK_V \frac{\partial K_V}{\partial V} = R\theta \frac{\partial K_V}{\partial \theta}, \tag{4.14}$$

as *a necessary and sufficient condition that an ideal gas having a constant, non-vanishing difference of specific heats shall obey the* LAPLACE-POISSON *Law along every adiabat.* The reader will easily verify that the second and third equations of (4.13) provide a solution of this equation. The status of (4.13) will be made precise just after the statement of Property 9 in Chapter 11.

Making use of the result stated in Remark 1, we can show that (4.14) is also *a necessary and sufficient condition that an ideal gas satisfying* MAYER's *Assumption shall obey the* LAPLACE-POISSON *Law in every adiabatic process.*

CHAPTER 5

Examples.

Since in the theory so far constructed the three constitutive functions ϖ, Λ_V, and K_V are independent of one another, a fluid body with a given pressure function ϖ may have as adiabats curves of any shape compatible with arbitrary assignment of the functions Λ_V and K_V satisfying the conditions in Axiom II. This generality is excessive. The pressure function ϖ of a body should lay conditions upon Λ_V and K_V and thus restrict the shapes allowable to the adiabats of that body. One of the objectives of thermodynamics is to find such conditions, and Chapter 9 will provide them. In considering now some possible forms of adiabats, isobars, piezotropes, and neutral curves, for economy's sake we choose only such as are compatible with all restrictions to be derived later on the basis of further axioms.

In all these examples the letter J will denote some positive constant having the dimensions of work per unit of heat.

EXAMPLE 1 (The ideal gas with constant ratio of specific heats). The adiabats of an ideal gas for which $\gamma = $ const. are given by (4.10). For the case in which $\gamma = \frac{5}{3}$, they are sketched in Figure 5. For completeness we list here a set of constitutive relations* compatible with the assumption that $\gamma = $ const.:

$$\mathscr{D}: \text{the entire } V\text{-}\theta \text{ quadrant.}$$

$$\varpi = R\theta/V. \qquad\qquad (2.9)_r$$

$$J\Lambda_V = \varpi. \qquad\qquad (3.13)_r$$

---

* Here and henceforth an r subscript to an equation number indicates that that equation is here repeated for the reader's convenience.

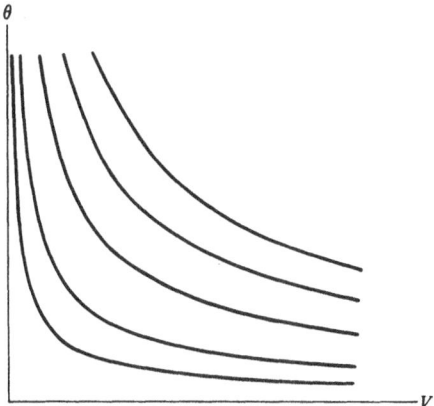

Figure 5. Adiabats of an ideal gas with constant ratio of specific heats (case drawn: $\gamma = \frac{5}{3}$).

$$JK_V = \text{const.} > 0. \tag{5.1}$$

$$\text{Adiabats: } \theta V^{\gamma-1} = \text{const.} \tag{4.10}_r$$

Adiabats and isotherms provide a co-ordinate system for all of $\mathscr{D}$.

EXAMPLE 2 (The Van der Waals fluid). For given positive constants R, a, and b

$$\mathscr{D}: V > b, \qquad \frac{R\theta}{(V-b)^2} > \frac{2a}{V^3}.$$

$$\varpi = \frac{R\theta}{V-b} - \frac{a}{V^2}. \tag{2.10}_r$$

$$J\Lambda_V = \varpi + \frac{a}{V^2}. \tag{5.2}$$

$$JK_V = \text{const.} > 0.$$

$$\text{Adiabats: } \theta(V - b)^{R/(JK_V)} = \text{const.} \tag{5.3}$$

Figure 6 shows for the special case in which $R/(JK_V) = \frac{2}{3}$ the adiabats in the v-τ plane of reduced variables introduced in Example 2, Chapter 2. These adiabats are just the same as those of an ideal gas for which $\gamma = \frac{5}{3}$, translated so as to make their vertical asymptote the line $v = \frac{1}{3}$. Adiabats and isotherms decussate throughout $\mathscr{D}$.

This particular Van der Waals fluid provides an example to show that from a given point (V, θ) it need not be possible to reach an arbitrarily low temperature by expansion along an adiabat.

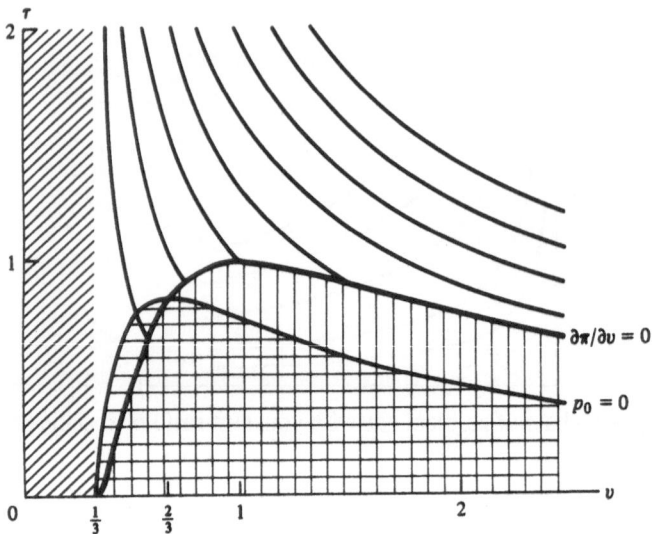

Figure 6. Possible adiabats of a Van der Waals fluid.

EXAMPLES 3 (Examples illustrating three kinds of piezotropes). In Figure 7 we see the adiabats and isobars of three particular fluids having a piezotrope that is also a neutral curve. For these fluids, a point is piezotropic if and only if it is neutral. For simplicity, each of the three cases shown in Figure 7 is chosen so as to make the piezotrope a straight line; for definiteness, the examples are specific ones, plotted from formulae which satisfy not only (2.2), the first inequality in (2.5), and (3.2) but also all the formal requirements of classical thermodynamics as determined below in Chapter 15. The examples here and in the following chapters may be regarded at pleasure as expressed in terms of special units or of reduced, dimensionless variables.

Figure 7A corresponds to the relations:

$$\mathscr{D}: \theta - \log \theta - V > 0.$$

$$\varpi = \theta - \log \theta - V.$$

$$J\Lambda_V = \theta - 1. \tag{A}$$

$$JK_V = V/\theta.$$

$$\text{Adiabats: } V(1 - 1/\theta) = \text{const.}$$

Any segment of the isotherm $\theta = 1$ in $\mathscr{D}$ is a piezotrope, a neutral curve, and an adiabat. Adiabats and isotherms decussate in the part of $\mathscr{D}$ where $\theta > 1$ and in the part where $\theta < 1$ but do not decussate throughout any part that contains a point at which $\theta = 1$.

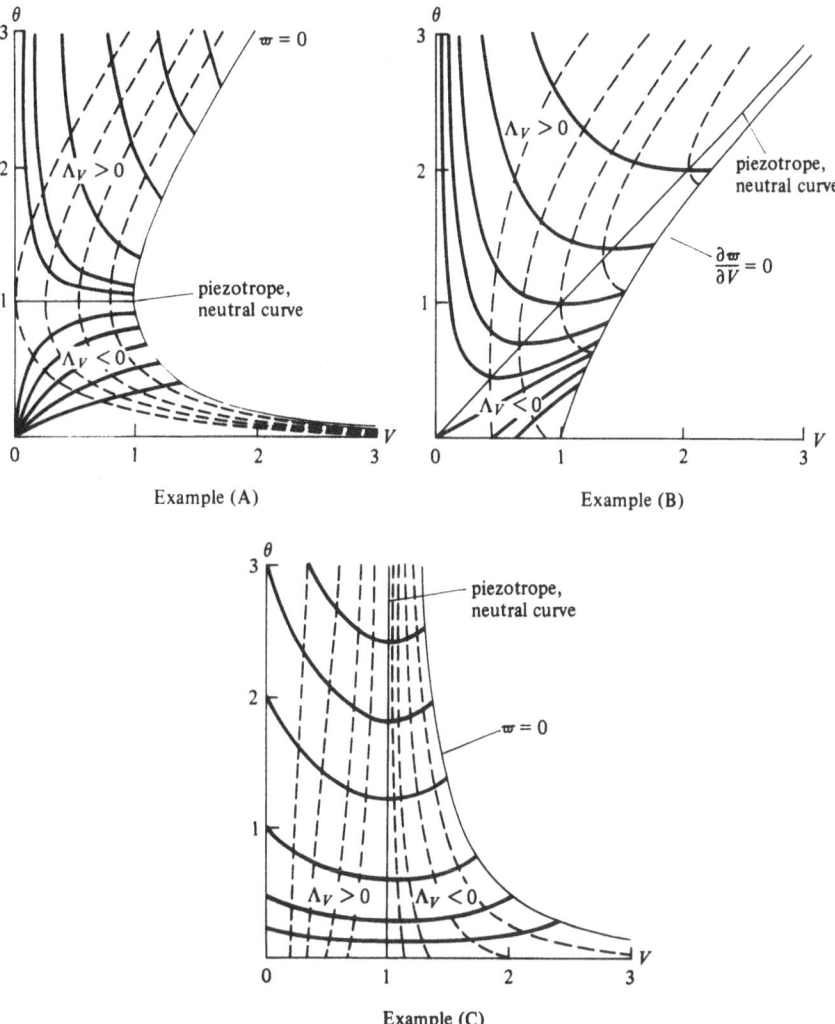

Figure 7. Various patterns of adiabats, isobars, piezotropes, and neutral curves. Heavy curves are adiabats. Dashed curves are isobars.

Figure 7B corresponds to the relations:

$$\mathscr{D}: \theta - V + \frac{1}{V^2} > 0.$$

$$\varpi = (\theta - V)^2 + \frac{2}{V}.$$

$$J\Lambda_V = 2(\theta - V)\theta.$$
$$JK_V = 2V\theta.$$

(B)

Adiabats: $\theta = \tfrac{1}{2}V + \dfrac{\text{const.}}{V}.$

Any segment of the ray $\theta = V$ is a piezotrope and a neutral curve. Adiabats and isotherms decussate in the part of $\mathscr{D}$ where $\theta < V$ and in the part where $\theta > V$ but do not decussate throughout any part that includes a point on the line $\theta = V$.

Figure 7C corresponds to the relations:

$$\mathscr{D}: \theta(1 - V) + \frac{1}{V} > 0.$$

$$\varpi = \theta(1 - V) + \frac{1}{V}.$$

$$J\Lambda_V = \theta(1 - V). \tag{C}$$

$$JK_V = \text{const.} > 0.$$

$$\text{Adiabats: } \theta \exp\left[\frac{V(1 - \frac{1}{2}V)}{JK_V}\right] = \text{const.}$$

Any segment of the isochor $V = 1$ is a piezotrope, a neutral curve, and an isobar. Adiabats and isotherms decussate in the part of $\mathscr{D}$ where $V < 1$ and in the part where $V > 1$, but they do not decussate throughout any part that includes a point at which $V = 1$.

We have seen that a neutral curve must be both an isothermal segment and an adiabat, or be neither. In the latter case, such adiabats as intersect the neutral curve must cross it. Example (A) illustrates the former possibility; Examples (B) and (C), the latter.

EXAMPLE 4 (BRIDGMAN's *"Hypothetical Liquid Water"*). Here would be the place to illustrate by an example from nature the form of the adiabats near a piezotropic point of water. To this end we should have to know the functions Λ_V and K_V at high pressures and low temperatures. Concerning them we find scant data, far from sufficient to guide a theorist even in some general direction.[1] The best we can do is turn to BRIDGMAN's "Hypothetical . . .

1. Ideally we should obtain Λ_V from direct experiments and then use CLAUSIUS' formula, the second equation of (15.15), below, so as to check the compatibility of thermodynamics with data on the thermal equation of state. We can find no record of any direct determination of Λ_V in the range of high pressures and low temperatures. The only course available seems to be to presume that thermodynamics does fit whatever data there be. In that case Λ_V is determined from the thermal equation of state by the thermodynamic relation expressed by the second equation of (15.15). As we have seen in Chapter 2, the nature of the thermal equation of state for water in the range of interest is a matter of dispute. Should we choose to shut our eyes to this problem, we should still need to find K_V. For it, as for Λ_V, we can find no record of direct determination. Usually K_p is measured, and then K_V is calculated from the thermodynamic formula appearing as the third equation of (15.15). Again, no test of thermodynamics can be made on the basis of data gotten by such procedure, for thermodynamics is presumed valid from the outset.

Collected data on specific heats are presented in Tables 113–120, pp. 257–264, of the book by DORSEY cited above in Footnote 6 to Chapter 2.

With Λ_V and K_V determined in this way, TAMMANN calculated the adiabat through 0°C at 1 atm. It is shown as Figure 18 on p. 134 of his book, which has been cited in Footnote 6 to Chapter 2. It has the form sketched here in Figure 8. Should we disregard BRIDGMAN's

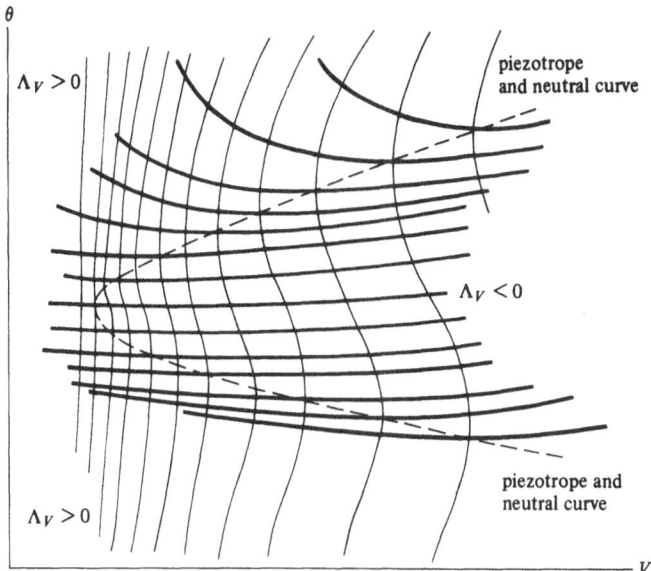

Figure 8. Possible arrangement of adiabats of BRIDGMAN's "Hypothetical... Liquid Water". Heavy curves are adiabats. Light curves are isobars.

Liquid Water", a possible set of isobars for which are shown in Figure 4. Figure 8 shows an arrangement consistent with the isobars shown in Figure 4 and with the idea, which we shall justify in Chapter 8, that generally $\Lambda_V > 0$ in parts of $\mathscr{D}$ where $\partial \varpi / \partial \theta > 0$, while $\Lambda_V < 0$ in parts where $\partial \varpi / \partial \theta < 0$. The adiabats and isotherms decussate in the regions lying on either side of the piezotrope but do not decussate at points on the piezotrope.

third table of data in favor of his first or second one, we should find those too scanty in values of ϖ at points near the piezotrope to permit numerical differentiation for Λ_V and subsequent determination of the adiabats by numerical integration of (4.3), even had we enough values of K_V.

KESTIN in §9.4 of his textbook, cited above in Footnote 9 to Chapter 2, states that the adiabats shown in his Figure 9.8(e) "have been drawn on the basis of empirically established thermodynamic properties of liquid water...". They are qualitatively like those we show in Figure 8.

In the end we have despaired of finding any empirical results that could serve as the basis of a thermodynamic theory of the anomalous behavior of water, not merely as an illustration of the way to apply such a theory.

CHAPTER 6

Emission–Absorption Estimates.

Historical Comment. A theorem of EULER asserts that Q has an integrating factor locally. That is, if (V, θ) is a point of $\mathcal{D}$, then in an open neighborhood of it there is a continuous, positive function f such as to make Q/f for an arbitrary process traversing points of that neighborhood the time derivative of a function $H_f(V, \theta)$. By (3.1)

$$\dot{H}_f = \frac{Q}{f} = \frac{1}{f}(\Lambda_V \dot{V} + K_V \dot{\theta}) \tag{6.1}$$

almost always. In the neighborhood where the integrating factor f exists, H_f has continuous partial derivatives, and

$$\frac{\Lambda_V}{f} = \frac{\partial H_f}{\partial V}, \qquad \frac{K_V}{f} = \frac{\partial H_f}{\partial \theta}. \tag{6.2}$$

A modern, precise statement and proof of EULER's theorem can be based on the theorem[1] concerning dependence of solutions on the initial data for the differential equation $d\theta/dV = -\Lambda_V/K_V$. If f has continuous partial derivatives, (6.2) shows that

$$\frac{\partial}{\partial \theta}\left(\frac{\Lambda_V}{f}\right) = \frac{\partial}{\partial V}\left(\frac{K_V}{f}\right). \tag{6.3}$$

Conversely, any local solution f of (6.3) with continuous partial derivatives gives rise to a function H_f such that (6.1) and (6.2) hold. REECH appealed to

1. See, for example, Problem 4 on p. 65 of I. G. PETROVSKI, *Ordinary Differential Equations*, New Jersey, Prentice-Hall, 1966.

this theorem so as to obtain some essential parts of his corpus of thermo-dynamic relations, but he need not have done so.

In this chapter we remark on some consequences that follow from (6.1) in conjunction with Axioms I and II. *We shall never use* EULER's *theorem directly.* Rather, in Chapter 9 we will prove that in what we shall define as a "thermodynamic part" of $\mathscr{D}$ there is a particular thermodynamic function of temperature alone that can be taken as f in (6.1), (6.2), and (6.3). Then the results obtained here will become applicable to thermodynamics.

Definition 13. The function H_f introduced above is a *pro-entropy* corresponding to the *integrating factor f.*

Remark 1. It is an axiom of the "caloric theory" of LAPLACE and CARNOT that 1 is an integrating factor for the entire domain $\mathscr{D}$. The corresponding pro-entropy H_1 is CARNOT's *heat function.* If LAPLACE's heat function is denoted by C, then $MC(\varpi(V, \theta), M/V) = H_1(V, \theta)$. By (6.3) we see that a necessary condition for the existence of a heat function is

$$\frac{\partial \Lambda_V}{\partial \theta} = \frac{\partial K_V}{\partial V}. \tag{6.4}$$

If a heat function H_1 exists, (6.2) reduces to

$$\Lambda_V = \frac{\partial H_1}{\partial V}, \qquad K_V = \frac{\partial H_1}{\partial \theta}, \tag{6.5}$$

formulae known, more or less, to the early thermodynamicists. If $\mathscr{D}$ is simply connected, the condition (6.4) is also sufficient that H_1 shall exist. In the caloric theory, obviously, $C^+ = C^-$ for every cyclic process.

Historical Scholion. The formula (6.4) seems first to have been remarked by KELVIN, and very late, in 1852. KELVIN then wrote[2] that from it "other remarkable conclusions...might have been drawn", so that "experimental tests might have been...suggested". Indeed. If K_V is a function of θ alone, (6.4) yields $\Lambda_V = f(V)$ locally. For an ideal gas we may apply (3.12) or (3.16) and so conclude that

$$K_p - K_V = \frac{Vf(V)}{\theta} \tag{6.6}$$

locally. Therefore, *for an ideal gas the caloric theory makes the following three statements equivalent locally:*

1. *Both K_p and K_V are functions of θ alone.*
2. $K_p - K_V = \alpha/\theta$, $\alpha = $ *const.*
3. $\Lambda_V = \alpha/V$, $\alpha = $ *const.*

2. W. THOMSON, "An account of Carnot's theory of the motive power of heat; with numerical results deduced from Regnault's experiments on steam", *Transactions of the Royal Society of Edinburgh* **16** (1849), 541–574 = [with annotations] *Papers* **1** (1882), 113–155.

A fortiori, *if both K_V and K_p are constants, they must be equal.* Thus *the caloric theory does not allow an ideal gas to have constant but unequal specific heats.*

In the early years of the nineteenth century LAPLACE argued that sonic oscillations were adiabatic, and POISSON supported this idea. Experiment had shown that the speed of sound in air was proportional to $\sqrt{\theta}$. As we have seen in Remark 2 after Corollary 4.1 in Chapter 4, if this fact is to square with the views of LAPLACE and POISSON when applied to an ideal gas, it is necessary that γ be constant. Then a second appeal to the same experimental facts shows that $\gamma > 1$. On the supposition that K_V is a function of θ only, it then follows by (6.6) that

$$K_p = \frac{Vf(V)}{(1 - \gamma^{-1})\theta} \qquad (6.7)$$

locally. No matter what be the theory of thermodynamics we choose to construct upon it, the caloric theory allows no other possibility of reconciling the simplest experiments on the speed of sound in air with the idea that sound waves are adiabatic.

We can find another objection to the caloric theory. Dropping the condition that γ be a constant, we can ask if within the caloric theory the LAPLACE–POISSON Law of Adiabatic Change (4.10) be compatible with MAYER's Assumption (3.14). In Remark 3 after Corollary 4.3 we have obtained the general condition (4.14). It remains only to see whether (4.14) and (6.4) be compatible with each other. By substituting (3.13) into (6.4) we obtain $JK_V = R \log V + k(\theta)$ locally. Since this formula for K_V does not satisfy (4.14), we conclude that *according to the caloric theory applied to ideal gases, the* LAPLACE–POISSON *Law of Adiabatic Change even with a ratio of specific heats that need not be constant is incompatible with* MAYER's *Assumption that the difference of specific heats is a positive constant.*

Although the foregoing arguments were within reach of the pioneers of thermodynamics, they gave no evidence of perceiving them, nor are those arguments yet well known in the critical literature.[3] Indeed, as we shall see in Chapter 15, in CARNOT's theory if $K_V = $ const., then K_p must decrease

3. The last of them makes its first appearance here. The others were presented in a lecture delivered by TRUESDELL on August 26, 1970, to a symposium at Aarhus on nineteenth-century mathematical physics. Numerous specialists on the work of CARNOT were present; some objected on principle to logical analysis of early science; but the ideas presented there seem to have gained some currency nevertheless. TRUESDELL's study in somewhat revised and expanded form was delivered as a series of three lectures at Udine in June, 1971, and has been published as *The Tragicomedy of Classical Thermodynamics* (1971), International Centre for Mechanical Sciences, Udine, Courses and Lectures, No. 70, Wien and New York, Springer-Verlag [1973], 41 pp. TRUESDELL takes this occasion to remark that the text of that pamphlet is no more than a preliminary extract from a larger study. In complying with the International Centre's request that he hand over to it the manuscript of his lectures there, he did not release that manuscript for general issue, let alone as a separate work for sale. The pamphlet was published without his consent or even knowledge beforehand. Although he believes most of the contents to be roughly correct, there are several statements he now would revise; many more sources should be analysed in a formal work with such a title; and assertions should be substantiated by detailed references.

when θ increases. Although today we know that gases do not behave in that way, and we should reject a formula like (6.7) at first glance, the dependence of the specific heats of real gases upon temperature had not been determined in the 1820s.

The reader of this scholion will easily see that CARNOT's effort, since it was based upon the caloric theory of heat, was doomed from the start, for no adjustment or application of that theory could lead to physically acceptable specific heats of ideal gases.

Remark 2. In 1854 CLAUSIUS, when he obtained (6.1) and (6.2) in his own way, gave reasons to believe that among the solutions f of (6.3) would be a function of θ alone, and he concluded that for all bodies that integrating factor could be taken as θ. The function H_θ that corresponds to this universal choice CLAUSIUS later called the *entropy*. We shall see in Chapter 15 that the existence of CLAUSIUS' entropy in a thermodynamic part of a constitutive domain follows mathematically from some simple assumptions about ideal gases.

Remark 3. For a given body, a process that traverses only points on which H_f has a constant value is called "isentropic". According to (6.1), *a process whose range lies in the domain of H_f is isentropic if and only if it is adiabatic.* In the ordinary presentations of thermodynamics the term "reversible" is applied to a theory in which this result holds. The theory developed here being of that kind, as indeed the Reversal Theorems witness, we have no need to introduce the term "isentropic" at all.

Remark 4. In the French literature one thermodynamic identity is commonly attributed to REECH. Ironically enough, it is one of the few formulae of the subject his memoir does not contain, although it does follow, if in excessively general form, from some of his results. To get it, we use (3.2), the second equation of (6.2), and the implicit function theorem to show that the equation $H = H_f(M/\rho, \theta)$ is locally invertible for θ:

$$\theta = \hat{\theta}(\rho, H). \tag{6.8}$$

If

$$\hat{\varpi}(\rho, H) \equiv \bar{\varpi}(\rho, \hat{\theta}(\rho, H)), \tag{6.9}$$

$\bar{\varpi}$ being the function introduced in Remark 2 after Corollary 4.1 in Chapter 4, then that theorem shows that

$$\frac{\partial \hat{\varpi}}{\partial \rho} = \gamma \frac{\partial \bar{\varpi}}{\partial \rho}, \tag{6.10}$$

which is "Reech's Theorem". We note that the identity (6.10), like Theorem 4, is not of thermodynamic origin, since no thermodynamic axiom has yet been introduced. However, to prove (6.10) we have used (6.2) and thus have appealed to EULER's theorem on the integrating factor.

All this is merely an excursion, for we shall have no occasion to use (6.10) anywhere in this tractate.

The following lemma and theorem refer to the quantities occurring in (1.5): C^+, C^-, $\mathscr{T}^+$, and $\mathscr{T}^-$. If $f(V, \theta)$ is a continuous function, for a given process V, θ we define $f(t)$ by $f(t) \equiv f(V(t), \theta(t))$. This function has a maximum $\max f$ and a minimum $\min f$ on $[t_1, t_2]$. If $\mathscr{T}^+$ is not empty, f has a finite supremum and a finite infimum on it. If $\mathscr{T}^-$ is not empty, f has a finite supremum and a finite infimum on it, too. All of the quantities $\max f$, $\min f$, $\sup_{\mathscr{T}^+} f$, $\inf_{\mathscr{T}^+} f$, $\sup_{\mathscr{T}^-} f$, $\inf_{\mathscr{T}^-} f$ are independent of the choice of process within a class of equivalent processes.

Lemma. *Let p and q be arbitrary constants. Then, for a cyclic process that traverses a set of points in the domain of an integrating factor f,*

$$qC^- = pC^+ + \int_{\mathscr{T}^+} Q\left(\frac{1}{f} - p\right) dt + \int_{\mathscr{T}^-} (-Q)\left(q - \frac{1}{f}\right) dt. \quad (6.11)$$

Proof. Integrating the first equation of (6.1) over $[t_1, t_2]$, we find that

$$\int_{t_1}^{t_2} \frac{Q}{f} \, dt = 0 \quad (6.12)$$

for the cyclic process. The identity (6.11) follows from (6.12), (1.4), and (1.5). $\square$

Theorem 5 (Emission–Absorption Estimates). *For a cyclic process that traverses a set of points in the domain of an integrating factor f,*

$$\frac{\max f}{\min f} C^+ \geq C^- \geq \frac{\min f}{\max f} C^+. \quad (6.13)$$

Equality subsists on the right-hand side if and only if $f = \max f$ on $\mathscr{T}^+$ and $f = \min f$ on $\mathscr{T}^-$; on the left-hand side if and only if $f = \min f$ on $\mathscr{T}^+$ and $f = \max f$ on $\mathscr{T}^-$. If $\mathscr{T}^+$ and $\mathscr{T}^-$ are not empty,

$$\frac{\sup_{\mathscr{T}^-} f}{\inf_{\mathscr{T}^+} f} C^+ \geq C^- \geq \frac{\inf_{\mathscr{T}^-} f}{\sup_{\mathscr{T}^+} f} C^+; \quad (6.14)$$

equality subsists on either side if and only if f is constant on both $\mathscr{T}^+$ and $\mathscr{T}^-$.

Proof. If we choose $p = 1/\max f$ and $q = 1/\min f$ in (6.11), the two integrands appearing in it are nonnegative. Then the second estimate of (6.13) follows at once. Equality subsists in the second estimate of (6.13) if and only if both the integrals in (6.11) vanish; that is, if and only if $f = \max f$ on $\mathscr{T}^+$ and $f = \min f$ on $\mathscr{T}^-$. To obtain the first estimate of (6.13), we exchange the choices of p and q. By making first the choice $p = 1/\sup_{\mathscr{T}^+} f$, $q = 1/\inf_{\mathscr{T}^-} f$, and then the choice $p = 1/\inf_{\mathscr{T}^+} f$, $q = 1/\sup_{\mathscr{T}^-} f$, we derive (6.14) and also the necessary and sufficient conditions for equality in (6.14). $\square$

Remark 1. The estimate (6.14) was derived only subject to the assumption that neither $\mathscr{T}^+$ nor $\mathscr{T}^-$ was empty. This restriction is of no account. From (6.13) we see at once that for a cyclic process, $C^- = 0$ if and only if $C^+ = 0$.

Equivalently, $\mathcal{T}^+$ is empty if and only if $\mathcal{T}^-$ is empty. Thus cyclic processes are of two kinds: the adiabatic ones, and those for which neither $\mathcal{T}^+$ nor $\mathcal{T}^-$ is empty.

Remark 2.   In the caloric theory $f = 1$, so equality subsists in (6.13). That is, $C^+ = C^-$ for all cyclic processes, as we have stated already in Remark 1 after Definition 13.

Remark 3.   Some authors regard the statement that a *cyclic process cannot absorb heat unless it emits some* as a law of thermodynamics. As Theorem 5 shows, this statement follows at once if Q has an integrating factor and thus does not necessarily presume or imply thermodynamic axioms of any kind. On the other hand, EULER's theorem delivers an integrating factor only locally, so the existence of an integrating factor throughout $\mathcal{D}$ might be regarded as having some physical content, and indeed Mr. SERRIN's theory, which is not restricted to bodies described by only two variables, derives from simple thermodynamic principles the existence of an integrating factor for a class of constitutive relations much more general than the one we consider in this tractate.

PART II

CARNOT'S GENERAL AXIOM

*Nothing in the whole range of Natural Philosophy is
more remarkable than the establishment of general laws
by such a process of reasoning.*

KELVIN, 1849

Carnot Processes. Carnot Cycles and Webs. Lemmas regarding Heat and Work for Cycles in a Carnot Web.

Definition 14. A cyclic process for which neither $\mathcal{T}^+$ nor $\mathcal{T}^-$ is empty is a *Carnot process* if there are constant temperatures θ^+ and θ^-, called the *operating temperatures* of the process, such that

$$\begin{aligned} \theta &= \theta^+ && \text{on } \mathcal{T}^+, \\ \theta &= \theta^- < \theta^+ && \text{on } \mathcal{T}^-. \end{aligned} \tag{7.1}$$

The cycle generated by a Carnot process is a *Carnot cycle*.

Remark 1. The concepts of Carnot process and Carnot cycle are constitutive ones. A cycle that is a Carnot cycle for one fluid body is not so, generally, for another.

Remark 2. Every process equivalent to a Carnot process is also a Carnot process with the same operating temperatures. The common operating temperatures of all processes in a Carnot cycle are the *operating temperatures* of that cycle.

Remark 3. The temperature θ^+, at which heat is absorbed, represents the temperature of the "furnace" or "boiler" of a heat engine; the lower temperature θ^-, at which heat is emitted, the temperature of the "refrigerator" or "condenser".

Figure 9 illustrates simple Carnot cycles of various kinds. The most common type is that shown as Example (A). This type is possible in a part of

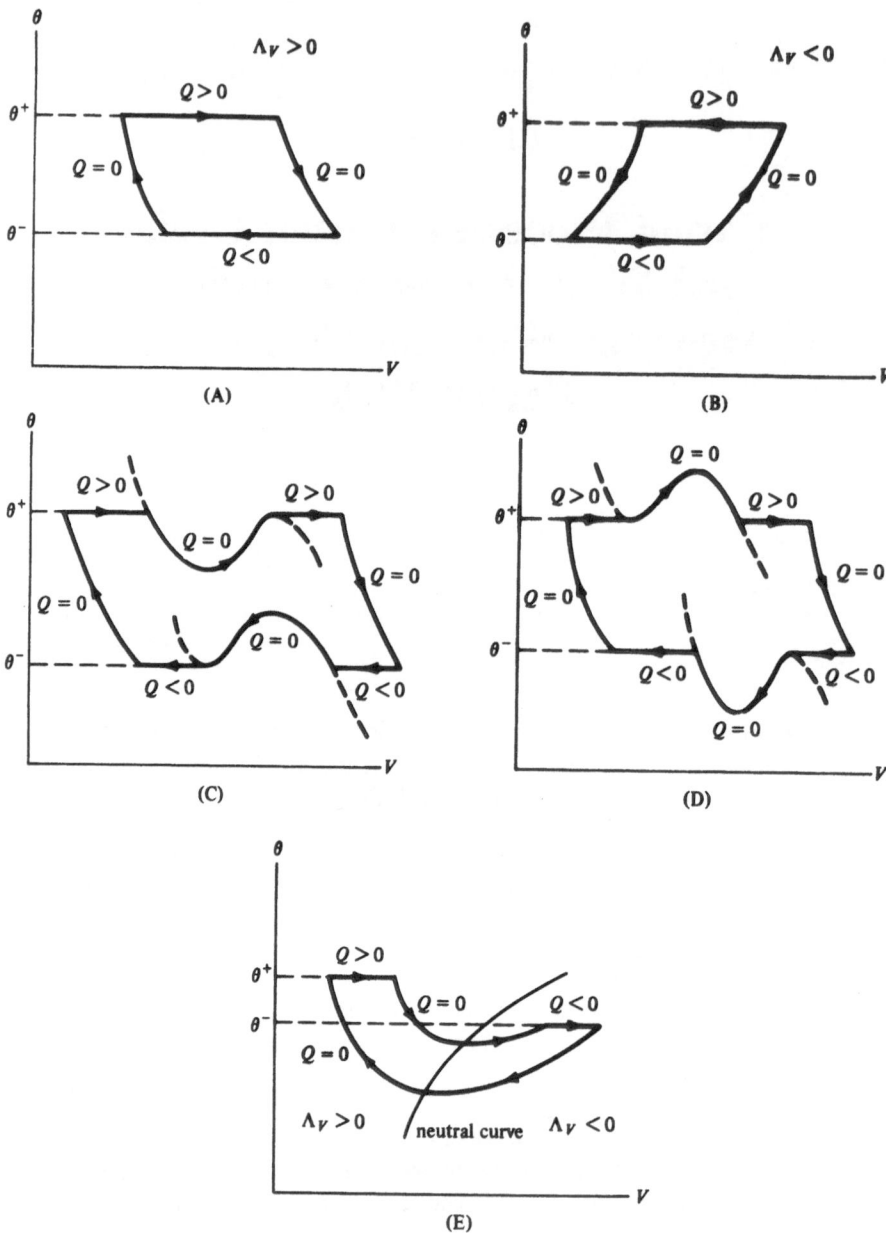

Figure 9. Examples of simple Carnot cycles. Sketches (A) and (B) represent ordinary Carnot cycles, while sketches (C), (D), and (E) represent simple Carnot cycles that are not ordinary.

$\mathscr{D}$ where $\Lambda_V > 0$; it is the only one illustrated in the works of discovery and in most modern textbooks. Example (B) is possible in a part where $\Lambda_V < 0$. The cycle in Example (A) has an orientation opposite to that of the cycle in

Example (B). Examples (C), (D), and (E) indicate possibilities for bodies with neutral points. Example (D) also shows that the operating temperatures need not coincide with the maximum and minimum temperatures of the processes in a Carnot cycle. All the cycles in Figure 9 are made up of adiabats and isothermal segments. Figure 10 shows simple cycles that are made up of adiabats and isothermal segments yet are not Carnot cycles.[1] In Example (A) of Figure 10 heat is absorbed and emitted at the same temperature while in Example (B) heat is absorbed at two different temperatures.

---

1. Textbooks along traditional lines are obscure in this regard. For example, in SOMMERFELD's book, cited above in Footnote 4 to Chapter 2, we find a brain twister:

> Exercise 1.6. Consider a Carnot cycle employing water between 2°C and 6°C with isothermal expansion at 6° and isothermal compression at 2°. Thus, provided the pressure be low enough, heat will be absorbed at both temperatures...and thus heat will be transformed entirely into work, in contradiction with the Second Law. How can this contradiction be resolved? To this end, sketch the adiabats and isotherms in the θ-V diagram near 4°C.
> Solution...Left of the minimum of the V-θ curve, the slope of the adiabats is positive, because $\partial V(p, \theta)/\partial\theta < 0$, but right of it negative because $\partial V(p, \theta)/\partial\theta > 0$, and at the minimum itself the direction of the adiabats is parallel to the V-axis. A sketch of the isobars in the θ-V diagram makes it easy to see that there are no adiabats that join the isotherms at 2° and 6°....

SOMMERFELD, MEIXNER, & BOPP provide neither the sketch of the adiabats which they require of the student who would solve their problem nor the sketch of the isobars they claim makes everything easy to see. It is, indeed, easy to provide by pure imagination various arrangements of adiabats in which a part where $\Lambda_V > 0$ is separated from a part where $\Lambda_V < 0$ by an asymptote, so that no adiabat can lead from one of those parts into the other. An example is shown above as Figure 7A in Chapter 5. However, if we accept the results of classical thermodynamics as presented, e.g., below in Chapter 15, in particular (15.15), then no such pattern would be compatible either with any of the experimentally determined isobars for water nor with the adiabat through 0° and 1 atm that TAMMANN obtained; these results suggest a pattern of the kind sketched in Figure 8 in Chapter 5. In that diagram there are adiabats connecting pairs of sufficiently near isotherms. One of these, indeed, will pass through the piezotropic point for 2° and connect two different volumes at 6°, as indicated in Figure 10A.

There is no contradiction here with what SOMMERFELD, MEIXNER, & BOPP call The Second Law, for heat is emitted in expansion on the portion of the isotherm at 6° that lies in the part where $\Lambda_V < 0$: *A cycle one of whose isotherms crosses a piezotrope is not a Carnot cycle!* SOMMERFELD himself (§6A) requires (if obscurely) that heat be absorbed all along the isotherm at higher temperature. His "contradiction" seems to arise from the treatment of the Carnot cycle in his own text (§6A), where he states that "the adiabats must have qualitatively the character of those for an ideal gas...." Near a piezotrope, such a claim becomes absurd. A possible Carnot cycle with operating temperatures 6° and 2° is sketched in Figure 9E; in it, the water suffers isothermal expansion at *both* those temperatures.

In its hidden assumptions and vagueness, SOMMERFELD's treatment is typical of those found in textbooks. Most of them, however, avoid trouble by failing to take up the thermodynamic aspects of the anomalous behavior of water. SOMMERFELD, MEIXNER, & BOPP seem to have seen the problem but, because part of the formal apparatus they provided is valid only in domains where $\Lambda_V > 0$, they were unable to solve it.

Figures 10 and 9E, which conform with "BRIDGMAN'S Hypothetical Liquid Water", show that the "contradiction" of SOMMERFELD, MEIXNER, & BOPP is like those that result when the student is told to begin by drawing a triangle with two right angles; their "solution" is like a claim that such a triangle cannot exist because no polygon has two interior right angles. The "contradiction" was explained correctly by J. S. THOMSEN & T. J. HARTKA in the paper cited in Footnote 10 to Chapter 2. The cycles themselves had been noticed and analysed correctly earlier by J. E. TREVOR, "Carnot cycles of unfamiliar types", *Sibley Journal of Engineering* 42 (1928), 274–278.

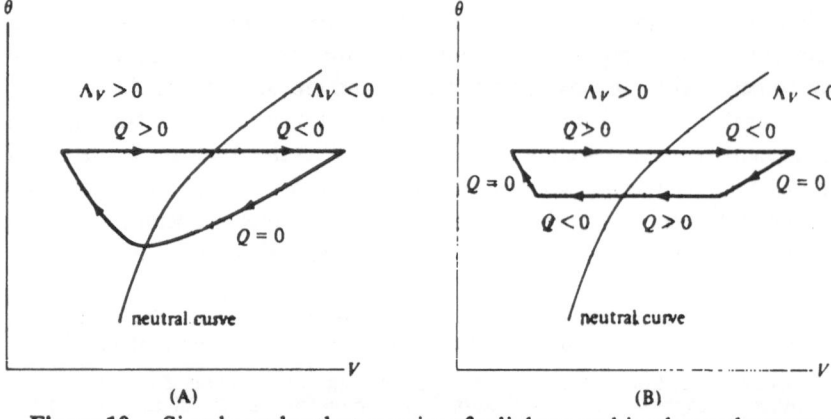

Figure 10. Simple cycles that consist of adiabats and isothermal segments yet are not Carnot cycles.

Remark 4. $\mathscr{T}^+$ has been defined as the set of t such that $Q(t) > 0$. Because Q is a piecewise continuous function, any such t lies in an interval on which $Q(t) > 0$. If $t = t_1$, or $t = t_2$, this interval need not be open. However, for a Carnot process $\theta = $ const. for all t such that $Q(t) > 0$, so any such t lies in an interval during which $\theta = $ const., and (3.1) shows that $\Lambda_V \dot{V} > 0$ on that interval. Thus, each time in $\mathscr{T}^+$ belongs to an interval on which one and only one of the following conditions holds:

$$\Lambda_V > 0 \quad \text{and} \quad \dot{V} > 0,$$
$$\Lambda_V < 0 \quad \text{and} \quad \dot{V} < 0. \tag{7.2}$$

Likewise every time in $\mathscr{T}^-$ belongs to an interval on which one and only one of the following holds:

$$\Lambda_V > 0 \quad \text{and} \quad \dot{V} < 0,$$
$$\Lambda_V < 0 \quad \text{and} \quad \dot{V} > 0. \tag{7.3}$$

Remark 5. The previous remark makes it plain that in a neutral part of $\mathscr{D}$ there are no Carnot cycles. Figure 7A provides examples of neutral points which cannot be included by any simple Carnot cycle.

Definition 15. A Carnot cycle is *ordinary* if

1. it is simple;
2. its interior is contained in $\mathscr{D}$; and
3. Λ_V is of one sign on it and within it.

Remark 1. All points within and on an ordinary Carnot cycle are ordinary points. An elementary though somewhat tedious geometrical argument based on (4.3) and Remark 4 after Definition 14 shows that an ordinary Carnot cycle has the shape and orientation indicated in Example (A) of Figure 9 if $\Lambda_V > 0$ on and within the cycle; the shape and orientation of Example (B) in Figure 9 if $\Lambda_V < 0$ on and within the cycle. Remark 2 after Definition 12 in Chapter 4 makes it clear that in a sufficiently small neighborhood of an ordinary point *it is always possible to construct ordinary Carnot cycles.*[2] In particular, if on the isotherm $\theta = B$ there is an ordinary point of $\mathcal{D}$, then there is an ordinary Carnot cycle whose operating temperatures C and A satisfy the inequality $A < B < C$. Ordinary Carnot cycles play an important role in some of our arguments in the following chapters.

Remark 2. From (2.15) we see that *a body undergoing a process of an ordinary Carnot cycle which lies in a part of $\mathcal{D}$ where $\Lambda_V > 0$ and $\partial \varpi / \partial \theta > 0$ or in a part where $\Lambda_V < 0$ and $\partial \varpi / \partial \theta < 0$ always does positive work.* Such a Carnot cycle serves as a model for the operation of a *motor* ("heat engine"). A motor represented by a cycle in a part of $\mathcal{D}$ in which $\Lambda_V > 0$ does work by expansion at the higher operating temperature and contraction at the lower one; in a part where $\Lambda_V < 0$, by expansion at the lower operating temperature and contraction at the higher one. On the other hand, a body undergoing a process of an ordinary Carnot cycle in a part where $\Lambda_V > 0$, $\partial \varpi / \partial \theta < 0$, or $\Lambda_V < 0$, $\partial \varpi / \partial \theta > 0$, would consume work rather than give it out. The Doctrine of Latent and Specific Heats allows these possibilities. That it does so, is one more evidence that it is not a thermodynamics but only a substructure for one.

Remark 3 (Motors and Coolers). The reverse of a Carnot process is not a Carnot process. If a body gives out work when it undergoes a Carnot process, in undergoing the reverse of that process it must suffer work to be done upon it. It absorbs heat at the lower operating temperature of the original Carnot process. The reverse process serves as a model for a *cooler*.

Remark 4 (Subdivision of Ordinary Carnot Cycles). Let an ordinary Carnot cycle be given. By subdividing one of the isothermal segments of the Carnot cycle at one of its interior points, then adjoining to it an adiabat that connects that isotherm to the other one, we may subdivide the given Carnot cycle into two smaller Carnot cycles. In a similar way, we may subdivide the given Carnot cycle at any interior point on one of its adiabats. More generally, as illustrated in Figure 11, within a given ordinary Carnot cycle we may construct infinitely many more. A strict justification of the construction is a consequence of the existence theory for the ordinary differential equation $dV/d\theta = -K_V/\Lambda_V$.

Suppose that $B > A$, and let the isotherms defining the cycle be $\theta = A$ and $\theta = B$; let the defining adiabats of the cycle have the equations $V = \phi_0(\theta)$

2. The construction is founded upon the local existence theory for the ordinary differential equation $dV/d\theta = -K_V/\Lambda_V$. The details are explained on pp. 74–76 of M. GOLOMB & M. SHANKS, *Elements of Ordinary Differential Equations*, New York etc., McGraw-Hill, 1965.

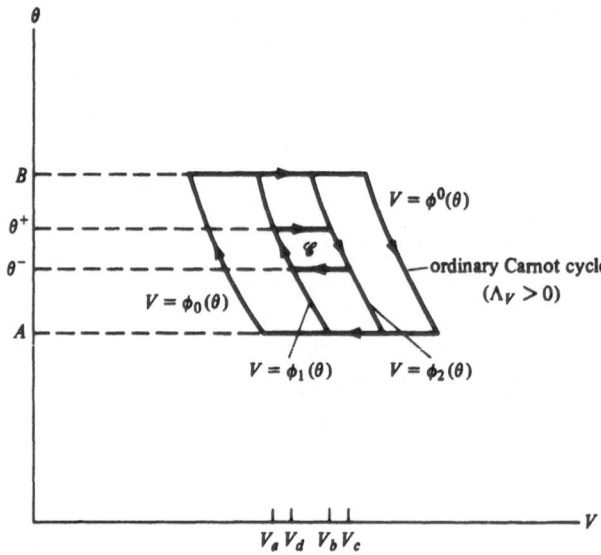

Figure 11. Description of a typical element of a Carnot web, drawn for the case in which $\Lambda_V > 0$.

and $V = \phi^0(\theta)$. In what follows, the numbers A and B and the functions ϕ_0 and ϕ^0 are fixed, being used only so as to define the part of $\mathcal{D}$ to which the analysis refers. We shall carry through the arguments on the supposition that $\Lambda_V > 0$ on and within the given ordinary Carnot cycle. Parallel arguments hold if $\Lambda_V < 0$.

Let θ^+ and θ^- be distinct intermediate temperatures: $B \geqq \theta^+ > \theta^- \geqq A$. Also, let $V = \phi_1(\theta)$ and $V = \phi_2(\theta)$ be the equations of two distinct intermediate adiabats: $\phi_0(\theta) \leqq \phi_1(\theta) < \phi_2(\theta) \leqq \phi^0(\theta)$ for every θ in $[A, B]$. The isotherms $\theta = \theta^+$, $\theta = \theta^-$ and the adiabats $V = \phi_1(\theta)$, $V = \phi_2(\theta)$ define an ordinary Carnot cycle $\mathscr{C}$ within the given ordinary Carnot cycle as shown in Figure 11. If the volumes V_a and V_b for a cycle such as $\mathscr{C}$ are defined by $V_a \equiv \phi_1(\theta^+)$, $V_b \equiv \phi_2(\theta^+)$, then any four numbers θ^+, θ^-, V_a, and V_b such that $A \leqq \theta^- < \theta^+ \leqq B$ and (V_a, θ^+) and (V_b, θ^+) are distinct points in the closed region bounded by the given ordinary Carnot cycle determine an ordinary Carnot cycle within it. Indeed, we may draw the isothermal segment from the point (V_a, θ^+) to the point (V_b, θ^+); then construct the adiabats through the two end points and prolong them until they reach the isotherm $\theta = \theta^-$; then complete the cycle by adjoining the segment they cut out of the isotherm $\theta = \theta^-$. While this construction is always possible within the given ordinary Carnot cycle, it may fail if we venture outside it, as is shown by the example of the Van der Waals fluid in Figure 6, Chapter 5. A glance at that figure reveals that the adiabats through the end points of an isothermal segment do not always reach the isotherms that correspond to sufficiently low temperatures.

Definition 16 (Carnot Webs). The collection of all ordinary Carnot cycles constructed like $\mathscr{C}$ in Figure 11 by subdivision of a given ordinary Carnot cycle, the operating temperatures of which are B and A, is called the *Carnot web* $\mathscr{W}_{A,B}$ *generated* by the given ordinary Carnot cycle. B and A are the *extreme temperatures* of the web.

Remark 1. If we have two Carnot webs with the same extreme temperatures, it may be possible to find a larger web that contains both of them. On the contrary, as a glance at Figure 6 in Chapter 5 shows, for a Van der Waals fluid two Carnot webs with subcritical extreme temperatures need not have any such common extension.

Remark 2. If on the isotherm $\theta = B$ there is an ordinary point of $\mathscr{D}$, Remarks 1 and 4 after Definition 15 show that if we choose A such as to make $B - A$ a sufficiently small positive number, then there is a Carnot web $\mathscr{W}_{A,B}$.

Remark 3. Given the point (V_a, θ^+) on some element of a Carnot web $\mathscr{W}_{A,B}$ as in Figure 11, we can construct a sequence of elements $\mathscr{C}$ of $\mathscr{W}_{A,B}$ which have only (V_a, θ^+) in common and whose enclosed areas approach 0. One way of doing so is to let $V_b \to V_a$ while θ^+ and θ^- are kept fixed in Figure 11. Corollary 2.2 in Chapter 3 shows that this way is equivalent to supposing that $C^+(\mathscr{C}) \to 0$.

We are now ready to state and prove two essential lemmas on Carnot webs.

Lemma 1. *Let a Carnot web $\mathscr{W}_{A,B}$ be given. Let $V = \phi_1(\theta)$ and $V = \phi_2(\theta)$ denote the equations of two fixed intermediate adiabats of the ordinary Carnot cycle generating the web, as indicated in Figure 11. Then there is a positive, continuously differentiable function $h_{\phi_1\phi_2}$ on $[A, B]$ such that if $\mathscr{C}$ is any element of $\mathscr{W}_{A,B}$ formed from the two fixed adiabats and two isotherms intermediate between $\theta = A$ and $\theta = B$,*

$$C^+(\mathscr{C}) = h_{\phi_1\phi_2}(\theta^+),$$
$$C^-(\mathscr{C}) = h_{\phi_1\phi_2}(\theta^-),$$

$$\text{(7.4)}$$

θ^+ and θ^- being the operating temperatures of $\mathscr{C}$.

Proof. If

$$h_{\phi_1\phi_2}(\theta) \equiv \int_{\phi_1(\theta)}^{\phi_2(\theta)} \Lambda_V(V, \theta)\, dV, \qquad \text{(7.5)}$$

then $h_{\phi_1\phi_2} > 0$; that $h_{\phi_1\phi_2}$ is continuously differentiable on $[A, B]$, is a consequence of (4.3) and the assumptions on Λ_V and K_V in Axiom II. Then (7.4) follows because of (3.7). □

Lemma 2. *Let a Carnot web* $\mathscr{W}_{A,B}$ *be given. There is a continuously differentiable function* $g_{\phi_1\phi_2}$ *on* $[A, B]$ *such that if* $\mathscr{C}$ *is any Carnot cycle of the type considered in Lemma 1,*

$$L(\mathscr{C}) = g_{\phi_1\phi_2}(\theta^+) - g_{\phi_1\phi_2}(\theta^-), \qquad (7.6)$$

θ^+ *and* θ^- *being the operating temperatures of* $\mathscr{C}$.

Proof. Throughout the closed region bounded by $\mathscr{C}$ the adiabats and isotherms decussate, so we may evaluate (2.15) by a repeated integral:

$$L(\mathscr{C}) = \int_{\theta^-}^{\theta^+} d\theta \int_{\phi_1(\theta)}^{\phi_2(\theta)} \frac{\partial \varpi}{\partial \theta} (V, \theta) \, dV \qquad (7.7)$$

Thus the desired function g is obtained if we set

$$g'_{\phi_1\phi_2}(\theta) \equiv \int_{\phi_1(\theta)}^{\phi_2(\theta)} \frac{\partial \varpi}{\partial \theta} (V, \theta) \, dV. \qquad (7.8)$$

That $g'_{\phi_1\phi_2}$ is continuous, follows because $\partial\varpi/\partial\theta$, ϕ_1, and ϕ_2 are continuous. $\square$

Remark. Starting with any element $\mathscr{C}$ of a Carnot web, by subdivision of $\mathscr{C}$ we may construct the Carnot web generated by $\mathscr{C}$. For adiabats and temperatures common to the two webs, the corresponding functions $g_{\phi_1\phi_2}$ and $h_{\phi_1\phi_2}$ are the same. However, as we noted in Remark 1 after Definition 16, for a given interval $[A, B]$ there may be two webs that are not both parts of any greater one. Although the interval $[A, B]$ is common to these webs, there need be no connection between the functions defined over them by Lemmas 1 and 2, as indeed the notation for them suggests.

Corollary. *A necessary and sufficient condition that the work done by every element of a Carnot web be positive is that for each choice of functions* ϕ_1 *and* ϕ_2 *representing intermediate adiabats the function* $g_{\phi_1\phi_2}$ *be an increasing function.*

The Doctrine of Latent and Specific Heats does not suffice to ensure that such be true. This fact is one more in evidence that the Doctrine of Latent and Specific Heats is not a thermodynamics but only the substructure for one.

Properties of Ideal Gases

Two important properties of ideal gases were demonstrated and recorded at the end of Chapter 3. To close this chapter, we state and prove two more. Both of these are numbered with subscript loc to indicate that while they are of a local character only, they will be extended later to statements of global validity.

Property 3_{loc}. Let the volumes V_a, V_b, V_c, and V_d for a typical element $\mathscr{C}$ of a Carnot web $\mathscr{W}_{A,B}$ be designated as in Figure 11. Then for a body of ideal gas, in the closed region bounded by the ordinary Carnot cycle generating the web the relation

$$\gamma = f(\theta) \tag{7.9}$$

holds if and only if

$$\frac{V_b}{V_a} = \frac{V_c}{V_d} \tag{7.10}$$

for every Carnot cycle in the web $\mathscr{W}_{A,B}$.

    Proof. Suppose that (7.10) be satisfied for every element of $\mathscr{W}_{A,B}$. Let θ^+ be a number in $]A, B]$. Fix two intermediate adiabats $V = \phi_1(\theta)$ and $V = \phi_2(\theta)$; consider all cycles in the web formed from them and having θ^+ as the higher of their operating temperatures (see Figure 11); and set $V_a \equiv \phi_1(\theta^+)$, $V_b \equiv \phi_2(\theta^+)$. Applying (7.10) to these cycles, we find that

$$\phi_2(\theta) = \frac{V_b}{V_a} \phi_1(\theta) \tag{7.11}$$

for θ in $[A, \theta^+]$. Differentiating (7.11) and using (4.3), we have

$$\frac{K_V}{\Lambda_V} \text{ at } (\phi_2(\theta), \theta) = \frac{V_b K_V}{V_a \Lambda_V} \text{ at } (\phi_1(\theta), \theta) \tag{7.12}$$

for every θ in $[A, \theta^+]$. Putting $\theta = \theta^+$, we show that

$$\frac{V\Lambda_V}{K_V} \text{ at } (V_b, \theta^+) = \frac{V\Lambda_V}{K_V} \text{ at } (V_a, \theta^+). \tag{7.13}$$

Since the points (V_a, θ^+) and (V_b, θ^+) can be chosen arbitrarily on the isothermal segment $\theta = \theta^+$ between the terminal adiabats $V = \phi_0(\theta)$ and $V = \phi^0(\theta)$, the function $V\Lambda_V/K_V$ is constant on every such isothermal segment for temperatures in $]A, B]$. That it is also constant on the isothermal segment with temperature A, follows by setting $\theta = A$ in (7.12). Therefore, $V\Lambda_V/K_V$ is a function of temperature alone in the closed region bounded by the ordinary Carnot cycle generating the web. For a body of ideal gas we have (3.16), so (7.9) follows.

    Conversely, suppose (7.9) holds for a body of ideal gas. Since Λ_V is of one sign within the region specified, by appeal to (3.16) we see that $\gamma - 1$ is also of one sign there. Applying (4.8) to an element of $\mathscr{W}_{A,B}$, we get (7.10). $\quad\square$

    Remark. The above proof also shows that Property 3_{loc} remains valid for all fluid bodies if in its statement the condition (7.9) is replaced by $V\Lambda_V/K_V = f(\theta)$.

    Historical Comment. That the condition $\gamma = \text{const.} > 1$ implies (7.10), was remarked, very late, by KELVIN (1853).

We may use Lemma 2 to infer a further property of ideal gases.

    Property 4_{loc}. Let a Carnot web $\mathscr{W}_{A,B}$ be given. Let ϕ_1 and ϕ_2 represent a pair of intermediate adiabats of the ordinary cycle generating $\mathscr{W}_{A,B}$. Then, for a body of ideal gas, in order that $g'_{\phi_1\phi_2}(\theta)$ shall depend on ϕ_1 and ϕ_2 alone, not on θ, it is necessary and sufficient that

$$\gamma = f(\theta) \tag{7.9}_r$$

in the closed region bounded by the ordinary cycle generating the web.

Proof. Substitution of (2.9) into (7.8) yields

$$g'_{\phi_1\phi_2}(\theta) = R \log \frac{\phi_2(\theta)}{\phi_1(\theta)}. \tag{7.14}$$

Property 4_{loo} then follows from Property 3_{loo}. $\square$

Remark. Let the volumes V_a, V_b, V_c, and V_d for a typical element of a Carnot web be designated as in Figure 11. Then (7.14) shows that for a body of ideal gas

$$\frac{\log \dfrac{V_c}{V_d}}{\log \dfrac{V_b}{V_a}} = \frac{g'_{\phi_1\phi_2}(\theta^-)}{g'_{\phi_1\phi_2}(\theta^+)}. \tag{7.15}$$

CHAPTER 8

Carnot's General Axiom. Local Theory: Reech's First Theorem, First Principal Lemma.

Axiom III (CARNOT's General Axiom, 1824). *For a given body, there is a function G such that if $\mathscr{C}$ is an ordinary Carnot cycle having θ^+ and θ^- as its operating temperatures, then*

$$L(\mathscr{C}) = G(\theta^+, \theta^-, C^+(\mathscr{C})) > 0. \qquad (8.1)$$

Clarification. The domain of the *Carnot function G* is the set of all triples θ^+, θ^-, $C^+(\mathscr{C})$ that can correspond to an ordinary Carnot cycle for the body in question. We do not need to specify this domain more precisely. Since every element of a Carnot web is an ordinary Carnot cycle, Axiom III applies to all the Carnot cycles in any Carnot web. Remark 2 after Definition 16 in Chapter 7 assures us that if the isotherm $\theta = B$ passes through an ordinary point of $\mathscr{D}$, then by choosing A such as to make $B - A$ a sufficiently small positive number we can construct a Carnot web $\mathscr{W}_{A,B}$. The elements of a single web suffice for the results we shall derive from CARNOT's Axiom in this chapter. In the succeeding chapter we shall use them to obtain relations valid in a much larger part of $\mathscr{D}$.

General Scholion. According to CARNOT,[1]

1. *Réflexions*, p. 16. See also p. 12. Later, on pp. 28–29, CARNOT stated "we may justly compare the motive power of heat with that of a fall of water: both have a maximum that cannot be exceeded, whatever be, on the one hand, the machine employed to receive the action of the water, on the other, the substance used to receive the action of the heat". He went on to suggest that the difference of temperatures applied to the working body of a heat engine was in some way analogous to the height through which water fell so as to become capable of driving a hydraulic machine. It is possible that this fragile analogy suggested to him his General Axiom.

Wherever a difference of temperature exists, there motive power can be produced.

From the context we learn that he claimed all Carnot cycles would produce motive power, that is, do positive work. Furthermore,[2]

The motive power of heat is independent of the agents employed to realize it; its quantity is determined solely by the temperatures of the bodies between which is effected, finally, the transport of the caloric.

Thus CARNOT in effect asserted also that the function G in (8.1) was a *universal function*, defined over all triples of positive numbers θ^+, θ^-, and $C^+(\mathscr{C})$ such that $\theta^+ > \theta^-$, and the same for all bodies. He was to be followed in this further claim by CLAUSIUS. Our Axiom III, which expresses only part of CARNOT's assumptions, allows G to be a constitutive quantity like $\mathscr{D}$, ϖ, Λ_V, and K_V.

CARNOT himself applied only special cases of Axiom III. His General Axiom formed the starting point of REECH's researches.

All the theorems in this chapter and the next six will derive from Axioms I–III and from them only. In Chapter 15 we shall impose the further axiom that G is the restriction of a universal function and determine the consequences of that axiom.

Remark. CARNOT's General Axiom asserts first of all that a given ordinary Carnot cycle produces a definite, positive amount of work. In the notation set forth on Figure 11, since any Carnot cycle $\mathscr{C}$ in a web is completely defined by the four quantities θ^+, θ^-, V_a, and V_b, any property of Carnot cycles in a web is a function of these four variables. The work done $L(\mathscr{C})$ and the heat absorbed $C^+(\mathscr{C})$, therefore, are the values of known functions of θ^+, θ^-, V_a, and V_b; (2.15), (4.3), and (3.7) show that these functions can be determined from the constitutive functions ϖ, Λ_V, and K_V. CARNOT's General Axiom asserts something more: The work done depends upon the two arguments V_a and V_b only through their effect upon $C^+(\mathscr{C})$. Referring back to (3.7), we may represent $C^+(\mathscr{C})$ as the area under the graph of $\Lambda_V(V, \theta^+)$ from V_a to V_b (Figure 12) and so express CARNOT's General Axiom as follows:

If two ordinary Carnot cycles with common operating temperatures θ^+ and θ^- have pairs of abscissae V_a, V_b and $V_{a'}$, $V_{b'}$ that correspond to equal areas under the graph of $\Lambda_V(V, \theta^+)$, those two cycles do the same work.

Historical Comment. CARNOT did not analyse or distinguish various kinds of Carnot cycles; his discussion seems to use only the Carnot cycles we here call "ordinary"; and, as we have stated in the Historical Comment following Corollary 2.2 in Chapter 3, he presumed always that $\Lambda_V > 0$. Our Axiom III

---

2. *Réflexions*, p. 38. CARNOT's term, "the fall of caloric", refers to the fact that in a heat engine a quantity of heat C^+ is carried, or, as KELVIN was to write later, "let down", from the higher temperature θ^+ to the lower temperature θ^-.

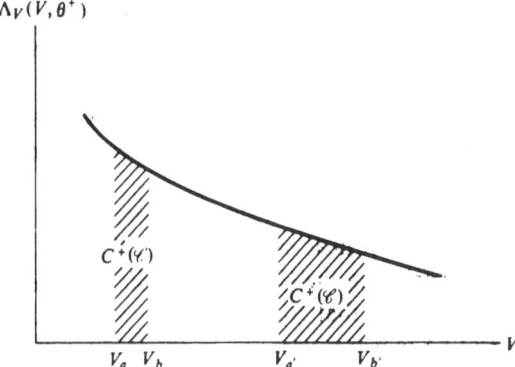

Figure 12. Heat absorbed on an isotherm. (The sketch is appropriate to the case in which $\Lambda_V > 0$.)

refers to ordinary Carnot cycles alone because these suffice for our proofs. Theorem 10, which we shall prove in Chapter 10, shows that the assertion of Axiom III holds for all Carnot cycles in a suitable part of $\mathscr{D}$, whether or not they be ordinary.

Lemma 1 (Constitutive Inequality). *In $\mathscr{D}$*

$$\Lambda_V \frac{\partial \varpi}{\partial \theta} \geqq 0. \tag{8.2}$$

Any piezotropic part of $\mathscr{D}$ is also a neutral part.

> *Proof.* To prove (8.2), it is sufficient to consider an ordinary point
> P. If $\Lambda_V > 0$ and $\partial \varpi / \partial \theta < 0$ at P, then in a neighborhood of P there
> are ordinary Carnot cycles oriented clockwise and including only
> points where $\partial \varpi / \partial \theta < 0$. From (2.15) we see that for these cycles
> $L(\mathscr{C}) < 0$, thus contradicting the inequality in CARNOT's General
> Axiom (8.1). Hence it is impossible that $\Lambda_V > 0$ and $\partial \varpi / \partial \theta < 0$ at P.
> A similar argument may be applied to a point at which $\Lambda_V < 0$ and
> $\partial \varpi / \partial \theta > 0$. Thus (8.2) is established.
>
> From (2.15) we see that a body in undergoing a process of a simple
> cycle whose interior lies in a piezotropic part of $\mathscr{D}$ does no work. If
> $\Lambda_V \neq 0$ at a point of such a part, then in a small enough neighborhood
> of that point we may construct ordinary Carnot cycles $\mathscr{C}$ for which
> $L(\mathscr{C}) = 0$, while CARNOT's General Axiom asserts that $L(\mathscr{C}) > 0$. We
> thus arrive at a contradiction, so the assumption that $\Lambda_V \neq 0$ at some
> point is false. Thus the second sentence of the lemma is proved. ☐

Remark 1. CARNOT's General Axiom cannot ensure the converse of the second sentence of Lemma 1, because no Carnot cycles exist in a neutral part of $\mathscr{D}$.

Remark 2. To prove Lemma 1, we have used only the inequality in CARNOT's General Axiom (8.1). The reader may easily show that, conversely, if the assertion of Lemma 1 is adjoined to Axioms I and II, then $L(\mathscr{C}) > 0$ for every ordinary Carnot cycle $\mathscr{C}$ in $\mathscr{D}$. Thus Lemma 1 expresses *a necessary and sufficient condition that all ordinary Carnot cycles do positive work.*

Historical Comment. CARNOT's General Axiom should be regarded as a summary of experience with heat engines, in which the working bodies are gases. As we have remarked in the Historical Comment after Corollary 2.2 in Chapter 3, CARNOT assumed that $\Lambda_V > 0$, so that work is done by expansion at the higher operating temperature, and in this presumption CLAUSIUS and ıll the other pioneers followed him. That water at atmospheric pressure attains its maximum density at about 4°C, had been known[3] for a century and a half before CARNOT's time; since $\partial \varpi / \partial \theta < 0$ on the part of the atmospheric isobar corresponding to temperatures below 4°C, (8.2) requires that $\Lambda_V \leqq 0$ on that part. As the creators of thermodynamics made no attempt to allow for this fact in their theories, they must not have thought it important. On the other hand, they all[4] claimed that their results held for "all bodies in nature— solids, liquids, or gases—".

Lemma 2. *Let a Carnot web $\mathscr{W}_{A,B}$ be given. Let x and y be fixed numbers in $[A, B]$ such that $x > y$. Then*

$$\lim_{z \to 0+} G(x, y, z) = 0. \tag{8.3}$$

Proof. Consider the collection of cycles in the web having the curve $V = \phi_0(\theta)$ (*cf.* Figure 11 in Chapter 7) as a common adiabat and having x and y as their operating temperatures. For every sufficiently small positive z, (3.7) shows that one of these cycles, say $\mathscr{C}$, is such that $C^+(\mathscr{C}) = z$. Thus $G(x, y, z)$ is defined for all sufficiently small positive z. Corollary 2.2 in Chapter 3 shows that as $z \to 0$, the area of the region inclosed by $\mathscr{C}$ tends to 0. By (2.15) we conclude that $L(\mathscr{C}) \to 0$. The lemma then follows from (8.1). □

Theorem 6 (REECH's First Theorem, 1851). *Let a Carnot web $\mathscr{W}_{A,B}$ be given, and let $\mathscr{S}$ be the set of pairs (x, y) such that $A \leqq y < x \leqq B$. On $\mathscr{S}$ there is a function F such that*

$$G(x, y, C^+(\mathscr{C})) = F(x, y)C^+(\mathscr{C}) \tag{8.4}$$

for every cycle $\mathscr{C}$ in $\mathscr{W}_{A,B}$ having x and y as its operating temperatures.

3. References are given by J. R. PARTINGTON, §VIII.C2 of *An Advanced Treatise on Physical Chemistry*, Volume 2, London *etc.*, Longmans, Green & Co., 1951. He particularly emphasizes the careful work of HOPE, published in 1805 in Scotland, England, and France, all three times in major journals.

4. *E.g.* E. CLAPEYRON in §V of "Mémoire sur la puissance motrice de la chaleur", *Journal de l'École Polytechnique* 14, Cahier 23 (1834), 153–190, variously translated into German and English.

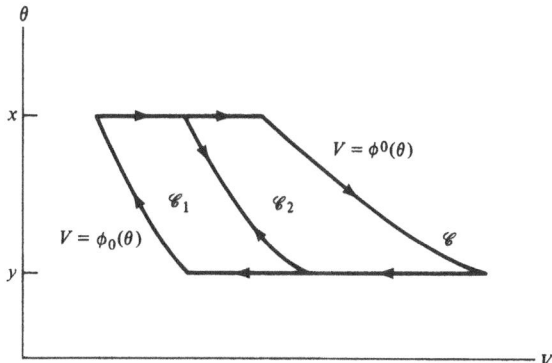

Figure 13. Construction for the proof of REECH's First Theorem (drawn for the case in which $\Lambda_V > 0$).

Proof. Let $\mathscr{C}$ be the element of $\mathscr{W}_{A,B}$ having x and y as its operating temperatures and having the curves $V = \phi_0(\theta)$ and $V = \phi^0(\theta)$ as its bounding adiabats (*cf.* Figure 11 in Chapter 7). We may subdivide $\mathscr{C}$ into the web elements $\mathscr{C}_1$ and $\mathscr{C}_2$ as shown in Figure 13. Because of (3.7)

$$C^+(\mathscr{C}) = C^+(\mathscr{C}_1) + C^+(\mathscr{C}_2). \qquad (8.5)$$

We next calculate the work done for processes in the cycles $\mathscr{C}$, $\mathscr{C}_1$, and $\mathscr{C}_2$, by use of (2.15) obtaining

$$L(\mathscr{C}) = L(\mathscr{C}_1) + L(\mathscr{C}_2). \qquad (8.6)$$

We now apply (8.1) to $\mathscr{C}$, $\mathscr{C}_1$, and $\mathscr{C}_2$. Substitution of the results into (8.6) followed by use of (8.5) yields a functional equation to be satisfied by G. Since we are holding x and y fixed, we may write $f(z) \equiv G(x, y, z)$. Then f satisfies

$$f(z + w) = f(z) + f(w) \qquad (8.7)$$

for all positive z and w such that $z + w = a \equiv C^+(\mathscr{C})$. However, by shortening the segment of the isotherm $\theta = x$ employed to obtain $\mathscr{C}$ we may obtain another Carnot cycle $\mathscr{C}^*$ belonging to $\mathscr{W}_{A,B}$ and having x and y as its operating temperatures. The argument we have applied to $\mathscr{C}$ applies equally to $\mathscr{C}^*$. Thus f must satisfy (8.7) for all positive z and w such that $z + w \leq a$. Moreover, (8.3) asserts that $f(z) \to 0$ as $z \to 0+$. It is well known[5] that such an f is linear. $\square$

---

5. First, we extend the definition of f from the domain $]0, a]$ to the whole real line in such a way as to satisfy (8.7) for all real numbers z and w. From (8.7) we see that in order to do so, we must set $f(0) \equiv 0$. For given positive a, any real number has a unique representation of the form $ma + t$, in which m is an integer and $0 \leq t < a$. Thus if

$$f(ma + t) \equiv mf(a) + f(t),$$

the f so defined extends the original f into a function defined over the whole line. An elementary but tedious analysis shows that f so extended does indeed satisfy (8.7) for all real numbers z and w.

Next we observe that since $f(z) \to 0$ as $z \to 0+$, f must be bounded in the open interval $]0, \alpha[$ for some sufficiently small positive number α. Any solution of (8.7) that is bounded

Both so as to assist those who would teach the subject to students untrained in rigorous analysis and also so as to uphold our claim that mathematics sufficient to construct a clear and clean thermodynamics was widely available at the time when the pioneers built the swamp of obscurity that thermodynamics became and has remained until now, we append here an elementary proof of the weak lemma that suffices for our purpose. The proof, for which we thank Mr. DAY, is close to that for functions defined over the whole real line which CAUCHY included in his *Analyse Algébrique*, published in 1821. In all the history of mathematics this textbook is one of the most widely read. Of the pioneers of thermodynamics from 1824 until 1872, KELVIN is the only one who showed evidence of mathematical education at the level of this textbook.

Lemma. *If f satisfies* (8.7) *for all positive z and w such that z + w is in* $]0, a]$, $a > 0$, *and if* $f(z) \to 0$ *as* $z \to 0+$, *then* $f(z) = f(a)z/a$ *in* $]0, a]$.

Proof. If $0 < z < a$ and if w is a sufficiently small positive number, then by (8.7)

$$f(z) + f(w) = f(z + w), \tag{A}$$

so

$$f(w) = f(z + w) - f(z)$$
$$= f(z - w + 2w) - f(z) = f(z - w) - f(z) + 2f(w). \tag{B}$$

This relation shows that if $f(w) \to 0$ as $w \to 0+$, then f is continuous at each z in $]0, a[$. It can also be shown that f is left-continuous at $z = a$. Moreover, from (A) we show that

$$f\left(\frac{a}{n}\right) = \frac{1}{n}f(a), \qquad n = 1, 2, 3, \ldots; \tag{C}$$

also, if $0 < z \leq a$ and if m is a positive integer such that $mz \leq a$, then (A) implies that

$$f(mz) = mf(z). \tag{D}$$

We take a fixed positive integer n, substitute a/n for z in (D), and so obtain

$$f\left(\frac{m}{n}a\right) = mf\left(\frac{a}{n}\right) = \frac{m}{n}f(a) \quad \text{if } 0 < \frac{m}{n} \leq 1. \tag{E}$$

Therefore f agrees with the function $[f(a)/a]z$ at the points $(m/n)a$. Since n may be as large as we like, this set of points is dense in $[0, a]$. Because f is continuous on $]0, a[$ and left-continuous at a, it can be no other function than $[f(a)/a]z$ in all of $]0, a]$. □

above on a set of positive Lebesgue measure is necessarily continuous. We do not need that fact here, however, since an elementary argument based upon (8.7) itself shows that if $f(z) \to 0$ as $z \to 0+$, then f is continuous. The celebrated theorem of CAUCHY asserts that every continuous solution of (8.7) on the whole real line is linear.

For an explanation of these matters and for references to the abundant literature on additive functions, the reader should consult the paper by JOHN M. BALL, "Measurability and continuity conditions for nonlinear evolutionary processes", *Proceedings of the American Mathematical Society* 55 (1976), 353–358.

Remark 1.   The function F whose existence is demonstrated in Theorem 6 is clearly unique.

Remark 2 (Efficiency).   Given a Carnot web $\mathscr{W}_{A,B}$, consider the collection of all cycles in the web that have fixed operating temperatures x_0 and y_0. For this collection the motive power is directly proportional to the heat absorbed, the constant of proportionality being $F(x_0, y_0)$. We can use the Carnot cycles of this collection to define the "mechanical equivalent" of a unit of heat as the motive power produced per unit of heat absorbed, namely $J \equiv F(x_0, y_0)$. The value of this "mechanical equivalent" depends upon the choice of x_0 and y_0 from $[A, B]$. If we now regard C^+ units of heat as equivalent to JC^+ units of work in the above sense, we can define the *efficiency* of any Carnot cycle $\mathscr{C}$ of $\mathscr{W}_{A,B}$ as the dimensionless ratio $F(x, y)/J$. Of course, $F(x, y)/J = L(\mathscr{C})/[JC^+(\mathscr{C})]$. The value of this efficiency depends on the choice of the temperatures x_0 and y_0 used to define J. Further, the efficiency of all Carnot cycles operating between x_0 and y_0 is 1. Clearly, the efficiency defined above is not absolute. Rather, it measures the performance of one engine in relation to the performance of another.

Remark 3.   Though it is possible to define the mechanical equivalent of a unit of heat as above, this clearly does not imply that a given amount of heat imparted to a given body can necessarily produce at most a certain amount of work. This much stronger requirement, the *uniform interconvertibility of heat and work* by a given body, is characterized below in Corollary 10.3 of Chapter 10. The still stronger requirement of *universal* as well as uniform interconvertibility will be taken up in Chapter 10 and in Chapter 15.

First Principal Lemma (Efficiency Theorem for Carnot Cycles in a Web). *Let $\mathscr{W}_{A,B}$ be a Carnot web, and let F be the function whose existence is demonstrated in Theorem 6. On $[A, B]$ there are continuously differentiable functions g and h such that $h(x) > 0$ and $(x - y)(g(x) - g(y)) > 0$ if $x \neq y$, and*

$$F(x, y) = \frac{g(x) - g(y)}{h(x)}. \tag{8.8}$$

The function h is unique to within a constant factor; when h has been determined, the function g is unique to within an additive constant.

Proof.   Writing x for θ^+ and y for θ^- in (7.6) and the first equation of (7.4) and using (8.4) (which we have obtained as a consequence of CARNOT's General Axiom), we see that

$$F(x, y) = \frac{g_{\phi_1\phi_2}(x) - g_{\phi_1\phi_2}(y)}{h_{\phi_1\phi_2}(x)}, \tag{8.9}$$

that $g_{\phi_1\phi_2}$ and $h_{\phi_1\phi_2}$ are continuously differentiable, and that $h_{\phi_1\phi_2}$ is positive. The left-hand side of (8.9) is independent of the choice of the adiabats represented by ϕ_1 and ϕ_2. Therefore, so is the right-hand side.

Referring to the notations set up in Figure 11 of Chapter 7, we may choose ϕ_1 as ϕ_0 and ϕ_2 as ϕ^0 if we like. The functions ϕ_0 and ϕ^0 are used only so as to specify the ordinary Carnot cycle that generates the Carnot web we are considering and hence are unaffected by the choice of the element $\mathscr{C}$. Writing g for $g_{\phi_0\phi^0}$ and h for $h_{\phi_0\phi^0}$, we obtain (8.8).The inequality in (8.1) requires that g be an increasing function and hence, since g is differentiable, that $g' > 0$ except upon a subset of $[A, B]$ with empty interior.

Now let $\bar{g}$ and $\bar{h}$ be any functions on $[A, B]$ such as to satisfy (8.8) when g is replaced by $\bar{g}$ and h by some positive function $\bar{h}$. Then if $B \geq x \geq y \geq A$,

$$\frac{\bar{g}(x) - \bar{g}(y)}{\bar{h}(x)} = \frac{g(x) - g(y)}{h(x)}. \tag{8.10}$$

Hence

$$\frac{\bar{g}(B) - \bar{g}(y)}{\bar{h}(B)} = \frac{g(B) - g(y)}{h(B)} \tag{8.11}$$

if $B \geq y \geq A$. Solving for $\bar{g}$, we find that

$$\bar{g}(y) = Kg(y) + N,$$

$$K \equiv \frac{\bar{h}(B)}{h(B)} > 0, \qquad N \equiv \bar{g}(B) - \frac{\bar{h}(B)}{h(B)} g(B) \tag{8.12}$$

if $A \leq y \leq B$. Putting the first equation of (8.12) back into (8.10) yields

$$K\frac{g(x) - g(y)}{\bar{h}(x)} = \frac{g(x) - g(y)}{h(x)} \tag{8.13}$$

if $B \geq x \geq y \geq A$. Therefore, since g is increasing,

$$\bar{h}(x) = Kh(x) \tag{8.14}$$

if $B \geq x > A$. If $\bar{h}$ is continuous, (8.14) holds also at $x = A$. Then the functions $\bar{g}$ and $\bar{h}$ have all the properties asserted for g and h, and any two functions $\bar{g}$ and $\bar{h}$ such as to satisfy the first equation of (8.12) and (8.14), K being a positive constant, may replace g and h in (8.8). $\square$

Remark 1. The last part of the proof delivers a result stronger than the one asserted by the theorem; namely, if $\bar{g}$ and $\bar{h}$ are any functions such as to satisfy the relation $F(x, y) = [\bar{g}(x) - \bar{g}(y)]/\bar{h}(x)$ when $B \geq x > y \geq A$, and if $\bar{h} > 0$, then $\bar{g}$ is given by the first equation of (8.12) on $[A, B]$, while $\bar{h}$ is given by (8.14) on $]A, B]$. Thus $\bar{g}$ is proved to be continuously differentiable on $[A, B]$, and $\bar{h}$ is proved to be continuously differentiable on $]A, B]$. In particular, since both (8.9) and (8.8) hold, $h_{\phi_1\phi_2} = K_{\phi_1\phi_2}h$, $K_{\phi_1\phi_2}$ being a positive constant that depends upon ϕ_1 and ϕ_2. It is clear that (8.10) by itself places no restriction upon $\bar{h}(A)$.

Remark 2. Recalling that g and h have been written for $g_{\phi_0\phi^0}$ and $h_{\phi_0\phi^0}$, we take note that the functions g and h so obtained depend upon the choice of the Carnot web $\mathscr{W}_{A,B}$, although the notation does not indicate this fact.

Corollary (Emission–Absorption Theorem for Carnot Cycles in a Web). *Let $\mathscr{C}$ be any element of a Carnot web $\mathscr{W}_{A,B}$. Then*

$$C^-(\mathscr{C}) = \frac{h(\theta^-)}{h(\theta^+)} C^+(\mathscr{C}), \tag{8.15}$$

θ^+ *and* θ^- *being the operating temperatures of* $\mathscr{C}$.

Proof. By (7.4) we obtain

$$C^-(\mathscr{C}) = \frac{h_{\phi_1\phi_2}(\theta^-)}{h_{\phi_1\phi_2}(\theta^+)} C^+(\mathscr{C}), \tag{8.16}$$

ϕ_1 and ϕ_2 representing the terminal adiabats of $\mathscr{C}$ (*cf.* Figure 11 in Chapter 7). As we have stated in Remark 1 after the First Principal Lemma, $h_{\phi_1\phi_2} = K_{\phi_1\phi_2}h$. $\square$

Subcorollary. *Let $\mathscr{C}$ be an element of a Carnot web $\mathscr{W}_{A,B}$, and let it have θ^+ and θ^- as its operating temperatures. Then*

$$L(\mathscr{C}) = \frac{g(\theta^+)}{h(\theta^+)} C^+(\mathscr{C}) - \frac{g(\theta^-)}{h(\theta^-)} C^-(\mathscr{C}). \tag{8.17}$$

Proof. It suffices to put (8.8) into (8.4) and then use (8.15). $\square$

Historical Comment. REECH's Second Theorem asserts that there is a function Γ such that if $\mathscr{C}$ is an element of a web, then

$$L(\mathscr{C}) = \Gamma(\theta^+)C^+(\mathscr{C}) - \Gamma(\theta^-)C^-(\mathscr{C}). \tag{8.18}$$

REECH's failure to analyse his function Γ and in particular to relate it to $C^+(\mathscr{C})$ and $C^-(\mathscr{C})$ through (8.15) resulted in his failure to obtain the specific results that we present in this tractate. REECH's function Γ depends upon some reference temperature; his conditions are merely necessary, not sufficient for CARNOT's General Axiom to be compatible with Axiom I and the Doctrine of Latent and Specific Heats; and most of his equations present a specious generality.

CHAPTER 9

Basic Constitutive Restrictions in a Thermodynamic Part and in the Normal Set. Pro-Entropy and Internal Pro-Energy.

Thus far we have dealt with Carnot cycles as ends in themselves. It is the singular merit of CARNOT to have seen that thermodynamics must relate the three constitutive functions ϖ, Λ_V, and K_V, thus far left arbitrary, and so must have far-reaching reflections upon the mechanical and caloric properties of the bodies it models. CARNOT, CLAPEYRON, and REECH, considering in effect only a simply connected $\mathcal{D}$ in which $\Lambda_V > 0$, derived relations connecting ϖ, Λ_V, and K_V. The restrictions CARNOT and CLAPEYRON obtained are valid only in the caloric theory; those derived by REECH are indeed necessary for the truth of Axiom III but, being excessively and illusorily general, are not sufficient for it.

Second Principal Lemma (Local Basic Constitutive Restrictions). *Let a Carnot web $\mathcal{W}_{A,B}$ be given. Let g and h be the functions on $[A, B]$ introduced in the First Principal Lemma. At any point of the closed region bounded by the ordinary Carnot cycle generating the web $\mathcal{W}_{A,B}$*

$$\frac{\partial}{\partial \theta}\left(\frac{\Lambda_V}{h}\right) - \frac{\partial}{\partial V}\left(\frac{K_V}{h}\right) = 0, \qquad (9.1)$$

$$\frac{g'}{h}\Lambda_V = \frac{\partial \varpi}{\partial \theta}. \qquad (9.2)$$

Proof. Let (V_a, θ^+) be a fixed interior point of the region of interest. We can always construct a Carnot cycle $\mathcal{C}$ belonging to the web and passing through this point, as labelled in Figure 11, Chapter 7. We may write (8.15) in the form

$$\left(1 - \frac{h(\theta^-)}{h(\theta^+)}\right) C^+(\mathscr{C}) = C^+(\mathscr{C}) - C^-(\mathscr{C}). \tag{9.3}$$

Using (3.7) and (3.6), we express this relation in terms of the constitutive functions Λ_V and K_V:

$$\left(1 - \frac{h(\theta^-)}{h(\theta^+)}\right) \int_{V_a}^{V_b} \Lambda_V(V, \theta^+)\, dV = \int_{\theta^-}^{\theta^+} d\theta \int_{\phi_1(\theta)}^{\phi_2(\theta)} \left(\frac{\partial \Lambda_V}{\partial \theta} - \frac{\partial K_V}{\partial V}\right) dV. \tag{9.4}$$

Clearly (9.4) is valid for all V_b and θ^- such as to make $V_b - V_a$ and $\theta^+ - \theta^-$ sufficiently small positive numbers, it being understood that the function ϕ_2, which represents the adiabat through (V_b, θ^+), depends on V_b. After differentiating (9.4) with respect to θ^- holding V_b fixed, we take the limit as $\theta^- \to \theta^+-$, and so show that

$$\frac{h'(\theta^+)}{h(\theta^+)} \int_{V_a}^{V_b} \Lambda_V(V, \theta^+)\, dV = \int_{V_a}^{V_b} \left[\frac{\partial \Lambda_V}{\partial \theta}(V, \theta^+) - \frac{\partial K_V}{\partial V}(V, \theta^+)\right] dV. \tag{9.5}$$

We differentiate this equation with respect to V_b, take the limit as $V_b \to V_a+$, and so show that

$$\frac{h'}{h} \Lambda_V = \frac{\partial \Lambda_V}{\partial \theta} - \frac{\partial K_V}{\partial V} \tag{9.6}$$

at the arbitrarily selected interior point of the closed region bounded by the ordinary Carnot cycle generating $\mathscr{W}_{A,B}$. Because the functions occurring in (9.6) are continuous, that relation subsists also on the boundary of this region. We show easily that (9.6) is equivalent to (9.1).

For the cycle $\mathscr{C}$ introduced above, application of (2.15), (3.7), and (8.8) yields

$$\int_{\theta^-}^{\theta^+} d\theta \int_{\phi_1(\theta)}^{\phi_2(\theta)} \frac{\partial \varpi}{\partial \theta}\, dV = \frac{g(\theta^+) - g(\theta^-)}{h(\theta^+)} \int_{V_a}^{V_b} \Lambda_V(V, \theta^+)\, dV. \tag{9.7}$$

We can arrive at (9.2) by following the procedure employed in deducing (9.6) from (9.4). $\quad\square$

Third Principal Lemma (Escape from the Web). *Let P and Q be any two ordinary points of $\mathscr{D}$ on the same isotherm. Then*

$$\frac{\dfrac{\partial \varpi}{\partial \theta}}{\Lambda_V} \text{ at } P = \frac{\dfrac{\partial \varpi}{\partial \theta}}{\Lambda_V} \text{ at } Q, \tag{9.8}$$

$$\frac{\dfrac{\partial \Lambda_V}{\partial \theta} - \dfrac{\partial K_V}{\partial V}}{\Lambda_V} \text{ at } P = \frac{\dfrac{\partial \Lambda_V}{\partial \theta} - \dfrac{\partial K_V}{\partial V}}{\Lambda_V} \text{ at } Q. \tag{9.9}$$

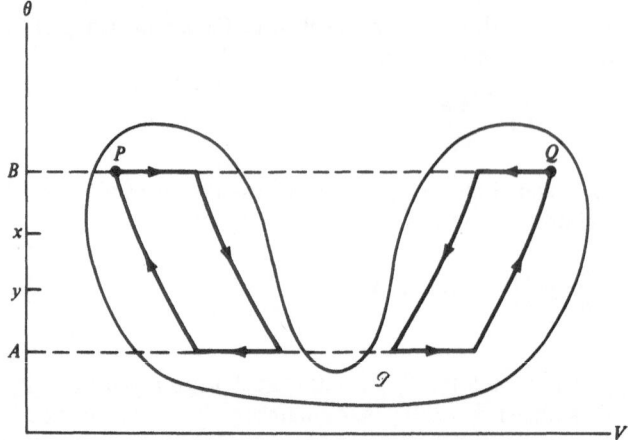

Figure 14. The sketch represents the case in which $\Lambda_V > 0$ at P and $\Lambda_V < 0$ at Q.

Proof. Let the isotherm be $\theta = B$. We can always construct ordinary Carnot cycles in the neighborhoods of P and Q with B and a lesser temperature A as their operating temperatures. Figure 14 has been drawn to illustrate such cycles when $\Lambda_V > 0$ at P and $\Lambda_V < 0$ at Q. The same argument works for the three other possible cases. Let the cycle near P generate the Carnot web $\mathscr{W}^P_{A,B}$; the cycle near Q, the web $\mathscr{W}^Q_{A,B}$. Let g and h be the functions on $[A, B]$ introduced in the First Principal Lemma for the web $\mathscr{W}^P_{A,B}$; let the corresponding functions for the web $\mathscr{W}^Q_{A,B}$ be $\bar{g}$ and $\bar{h}$. If x and y are two temperatures such that $B \geq x > y \geq A$, from (3.7) it is clear that we can select two Carnot cycles with those same operating temperatures, one cycle from the web $\mathscr{W}^P_{A,B}$ and the other from $\mathscr{W}^Q_{A,B}$, such that the same heat is absorbed in processes of both these cycles. CARNOT's General Axiom (8.1) asserts that the processes of these two cycles do equal work. Use of (8.8) shows that

$$\frac{g(x) - g(y)}{h(x)} = \frac{\bar{g}(x) - \bar{g}(y)}{\bar{h}(x)} \qquad (9.10)$$

for all x and y such that $B \geq x \geq y \geq A$. The argument after (8.10) shows that there are a positive constant K and a constant N such that

$$\bar{g}(x) = Kg(x) + N,$$
$$\bar{h}(x) = Kh(x), \qquad (9.11)$$

for all x in $[A, B]$. Hence, in particular

$$\frac{\bar{g}'(B)}{\bar{h}(B)} = \frac{g'(B)}{h(B)},$$
$$\frac{\bar{h}'(B)}{\bar{h}(B)} = \frac{h'(B)}{h(B)}. \qquad (9.12)$$

Applying (9.2) and (9.6) at P and Q and using (9.12), we arrive at (9.8) and (9.9). □

Remark. The main results of Chapter 8, namely, Theorem 6, the First Principal Lemma, its Corollary, and the Second Principal Lemma, were proveu by applying Axiom III to the elements of a single Carnot web. In proving the Third Principal Lemma we have applied CARNOT's Axiom to the cycles belonging to two different webs. Nevertheless the results remain local, for we have not yet shown that g and h exist over an interval greater than $[A, B]$, which may be very small.

We are now ready to define a much larger part of $\mathscr{D}$ and prove that the restrictions (9.1) and (9.2) hold throughout it.

Definition 17. A point of $\mathscr{D}$ is a *thermodynamic point* if each of its neighborhoods contains an ordinary point. A nonempty open set of thermodynamic points is a *thermodynamic part* $\mathscr{D}_{\text{th}}$ of $\mathscr{D}$ if also

1. it is simply connected; and
2. each isotherm that intersects it does so at at least one ordinary point.

A process that traverses only points of $\mathscr{D}_{\text{th}}$ is a *process in* $\mathscr{D}_{\text{th}}$.

Remark 1. Every ordinary point is a thermodynamic point. A point is thermodynamic if and only if it is the limit of a sequence of ordinary points. A thermodynamic point may be a neutral point. The points of $\mathscr{D}$ are of two mutually exclusive kinds: points which belong to a neutral part, and thermodynamic points.

Remark 2. A neutral part is never a thermodynamic part. A thermodynamic part cannot contain a neutral part. From Lemma 1 of Chapter 8 we see that *a thermodynamic part cannot contain a piezotropic part*.

Remark 3. Every nonempty open subset of $\mathscr{D}$ that is not a neutral part consists of thermodynamic points. Such a subset need not be a thermodynamic part, however. For example, suppose the part of $\mathscr{D}$ that is not neutral is an open annulus of ordinary points; because it is not simply connected, it cannot be a thermodynamic part, although of course it is the union of two thermodynamic parts. A second example is provided by a simply connected constitutive domain $\mathscr{D}$ all of whose points are thermodynamic, but which contains an isothermal segment of neutral points. This possibility is illustrated in Example (A) of Figure 7 in Chapter 5. In Examples (B) and (C) of that figure $\mathscr{D}$ is a thermodynamic part. The same is true of Figure 8.

Since $\mathscr{D}_{\text{th}}$ is connected and open, the range of temperatures in it is an open interval. We shall denote this interval by $\mathscr{I}$.

Theorem 7 (Basic Constitutive Restrictions in $\mathcal{D}_{\text{th}}$). *Let $\mathcal{D}_{\text{th}}$ be a thermodynamic part of $\mathcal{D}$; let its interval of temperatures be $\mathcal{I}$. On $\mathcal{I}$ there are continuously differentiable functions g and h, the former being an increasing function and the latter a positive one, such that at every point of $\mathcal{D}_{\text{th}}$*

$$\frac{\partial}{\partial\theta}\left(\frac{\Lambda_V}{h}\right) - \frac{\partial}{\partial V}\left(\frac{K_V}{h}\right) = 0, \tag{9.1}_r$$

$$\frac{g'}{h}\Lambda_V = \frac{\partial\varpi}{\partial\theta}. \tag{9.2}_r$$

The function h is unique to within a constant factor; when h has been determined, g is unique to within an additive constant.

Proof. Let θ_0 be a temperature in $\mathcal{I}$. Then there is an ordinary point P of $\mathcal{D}_{\text{th}}$ on the isotherm $\theta = \theta_0$. The ratio $(\partial\varpi/\partial\theta)/\Lambda_V$ has a value at P. Moreover, the Third Principal Lemma assures us that $(\partial\varpi/\partial\theta)/\Lambda_V$ has the same value at every other ordinary point of $\mathcal{D}_{\text{th}}$ on the isotherm $\theta = \theta_0$. Since θ_0 is any temperature in $\mathcal{I}$, the foregoing observation enables us to define a function f on $\mathcal{I}$, the value $f(\theta_0)$ being $(\partial\varpi/\partial\theta)/\Lambda_V$ at any and every ordinary point of $\mathcal{D}_{\text{th}}$ on the isotherm $\theta = \theta_0$. The same reasoning applies to $(\partial\Lambda_V/\partial\theta - \partial K_V/\partial V)/\Lambda_V$ and so defines a function ϕ on $\mathcal{I}$. We have thus constructed functions f and ϕ on $\mathcal{I}$ such that at every ordinary point of $\mathcal{D}_{\text{th}}$

$$\frac{\dfrac{\partial\varpi}{\partial\theta}}{\Lambda_V} = f, \qquad \frac{\dfrac{\partial\Lambda_V}{\partial\theta} - \dfrac{\partial K_V}{\partial V}}{\Lambda_V} = \phi. \tag{9.13}$$

Since $\mathcal{D}_{\text{th}}$ is open and Λ_V is continuous, the above-selected ordinary point P lies in a neighborhood $\mathcal{N}$ consisting of ordinary points of $\mathcal{D}_{\text{th}}$. The left-hand sides of the first equation of (9.13) and the second equation of (9.13) are continuous functions in $\mathcal{N}$. Because they equal functions of θ alone there, those functions must be continuous at θ_0. Thus f and ϕ are continuous on $\mathcal{I}$.

From (8.2) we know that at an ordinary point Λ_V and $\partial\varpi/\partial\theta$ cannot be of opposite sign. By the first equation of (9.13) we conclude that

$$f \geqq 0. \tag{9.14}$$

Suppose now that $f(\theta) = 0$ on an open subinterval of $\mathcal{I}$; let θ_0 be any point of that subinterval; let P be an ordinary point of $\mathcal{D}_{\text{th}}$ on the isotherm $\theta = \theta_0$; and let $\mathcal{N}$ be an open neighborhood of P consisting of ordinary points of $\mathcal{D}_{\text{th}}$. Referring back to the first equation of (9.13), we see that we may select $\mathcal{N}$ so that $\partial\varpi/\partial\theta = 0$ throughout it. That is, $\mathcal{N}$ is a piezotropic part of $\mathcal{D}_{\text{th}}$. As we have stated in Remark 2 after Definition 17, no such $\mathcal{N}$ exists. Thus the assumption that f vanishes on an open interval is false. Therefore, $f(\theta) > 0$ except on a set having empty interior.

We may write (9.13) as follows: At every ordinary point of $\mathcal{D}_{\text{th}}$

$$\frac{\partial\Lambda_V}{\partial\theta} - \frac{\partial K_V}{\partial V} = \phi\Lambda_V, \qquad \frac{\partial\varpi}{\partial\theta} = f\Lambda_V. \tag{9.15}$$

The definition of $\mathscr{D}_{th}$ makes it clear that every point of $\mathscr{D}_{th}$ is the limit of a sequence of ordinary points of $\mathscr{D}_{th}$ at which (9.15) holds. Since the functions occurring in (9.15) are continuous, those relations subsist at every point of $\mathscr{D}_{th}$.

Now we define the functions $g(\theta)$ and $h(\theta)$ on $\mathscr{I}$ as follows:

$$h(\theta) \equiv \exp \int_{\theta_0}^{\theta} \phi(u) \, du,$$

$$g'(\theta) \equiv h(\theta)f(\theta),$$

(9.16)

θ_0 being some fixed temperature in $\mathscr{I}$. The definitions of g and h make them continuously differentiable and make h positive. It was shown after (9.14) that $f(\theta) > 0$ except on a subset of $\mathscr{I}$ with empty interior. From the second equation of (9.16) we conclude that g' also has this property. The mean-value theorem then shows that g is an increasing function on $\mathscr{I}$. Using (9.16) in (9.15), we show that (9.1) and (9.2) are satisfied at every point of $\mathscr{D}_{th}$.

Let $\bar{g}$ and $\bar{h}$ be differentiable functions on $\mathscr{I}$ such as to replace g and h in (9.1) and (9.2), $\bar{h}$ being positive. Then, by the result just established, $\bar{g}'/\bar{h} = g'/h$, and $\bar{h}'/\bar{h} = h'/h$. These relations imply that there are constants K and N such that $K > 0$ and for every θ in $\mathscr{I}$

$$\bar{g}(\theta) = Kg(\theta) + N,$$

$$\bar{h}(\theta) = Kh(\theta).$$

(9.17)

Further, $\bar{g}$ and $\bar{h}$ have all the properties asserted for g and h, and any two functions $\bar{g}$ and $\bar{h}$ satisfying (9.17) for some positive K can replace g and h in (9.1) and (9.2). $\quad\square$

Remark 1. In the last part of the above proof, $\bar{g}$ and $\bar{h}$ were assumed to be differentiable. That their derivatives are continuous is immediate from (9.17). Thus the proof delivers a result stronger than the one asserted by Theorem 7 (*cf.* also Remark 1 after the First Principal Lemma in Chapter 8).

Remark 2. Consider a Carnot web $\mathscr{W}_{A,B}$ generated by an ordinary Carnot cycle in $\mathscr{D}_{th}$. Theorem 7 extends the results of the Second Principal Lemma to the entire domain $\mathscr{D}_{th}$, in which Λ_V need not be of one sign. In this sense the functions g and h on $\mathscr{I}$ in Theorem 7 are extensions of the corresponding functions on $[A, B]$ appearing in the Second Principal Lemma. Indeed, the definition of $\mathscr{D}_{th}$ was so framed as to provide neat extensions. In the proof of Theorem 7 we used the fact that $\mathscr{D}_{th}$ had been assumed connected; that it is assumed simply connected becomes important only later.

Remark 3. At this stage the functions g and h are clearly constitutive. In Chapter 15 we shall lay down further axioms such as to make these functions *common to all bodies* and to give them explicit *universal forms*. Our theory will thereby reduce to the theory of CLAUSIUS. Until then we shall continue to explore the consequences of Axioms I–III alone. By so doing we shall accumulate a treasury of theorems and formulae which can be specialized at will to Clausius' theory or to Carnot's. Thus we shall make it trivial to answer, once and for all, any question that may be asked as to how far the two theories agree or disagree with each other.

We now record some easy but important consequences of Theorem 7.

Corollary 7.1. *In $\mathscr{D}_{\text{th}}$*

$$\frac{\partial \varpi}{\partial \theta} = \frac{\partial}{\partial \theta}\left(\frac{g}{h}\Lambda_V\right) - \frac{\partial}{\partial V}\left(\frac{g}{h}K_V\right),$$

$$h'\frac{\partial \varpi}{\partial \theta} = g'\left(\frac{\partial \Lambda_V}{\partial \theta} - \frac{\partial K_V}{\partial V}\right). \tag{9.18}$$

Corollary 7.2. *Every neutral point of $\mathscr{D}_{\text{th}}$ is a piezotropic point. Let θ_0 be a temperature in $\mathscr{I}$. If $g'(\theta_0) = 0$, all points of $\mathscr{D}_{\text{th}}$ on the isotherm $\theta = \theta_0$ are piezotropic points; if $g'(\theta_0) \neq 0$, a point of $\mathscr{D}_{\text{th}}$ on the isotherm $\theta = \theta_0$ is piezotropic if and only if it is neutral.*

Corollary 7.3. *The partial derivative $\partial^2 \varpi / \partial V \partial \theta$ exists and is continuous throughout $\mathscr{D}_{\text{th}}$. If $g''(\theta_0)$ exists at a temperature θ_0 in $\mathscr{I}$, then $\partial^2 \varpi / \partial \theta^2$ exists at every point of $\mathscr{D}_{\text{th}}$ on the isotherm $\theta = \theta_0$.*

Corollary 7.4. *If g'' exists throughout $\mathscr{I}$, then*

$$\frac{g'^2}{h}\frac{\partial K_V}{\partial V} = g'\frac{\partial^2 \varpi}{\partial \theta^2} - g''\frac{\partial \varpi}{\partial \theta} \tag{9.19}$$

at every point of $\mathscr{D}_{\text{th}}$. If $\mathscr{D}_{\text{th}}$ is also isothermally convex,[1]

$$\frac{g'^2}{h}K_V = \int \left[g'\frac{\partial^2 \varpi}{\partial \theta^2} - g''\frac{\partial \varpi}{\partial \theta}\right]dV + K, \tag{9.20}$$

K being a function of θ alone.

    Proof. To prove (9.19), we need only multiply (9.1) by g'^2 and use (9.2). Integration delivers (9.20). □

---

    1. A subset of $\mathscr{D}$ is said to be *isothermally convex* if whenever it contains two distinct points on the same isotherm it contains also the isothermal segment that joins them. If f is a function defined on $\mathscr{D}$ such that $\partial f / \partial V = 0$, then f is constant upon each segment $\theta = $ const. in $\mathscr{D}$. If $\mathscr{D}$ is not isothermally convex, there is an isotherm $\theta = \theta_0$ that intersects $\mathscr{D}$ in at least two disjoint segments. Then f is constant on each of them, but we cannot conclude that the two constants are the same.

    A glance at Figure 2 in Chapter 2 shows that $\mathscr{D}$ for a Van der Waals fluid is not isothermally convex. We may apply (9.20) at once to such a fluid in the supercritical domain for the temperature or the pressure. To apply it in the subcritical domain, say for the pressure, we must first divide that domain into two parts, one appropriate to the gaseous phase and the other to the liquid. These two parts of $\mathscr{D}$ are isothermally convex. We obtain (9.20) with one choice of K in the former, another choice in the latter. A similar conclusion is valid in the subcritical domain for the temperature.

Remark 1. Consider an isothermally convex thermodynamic part $\mathscr{D}_{\text{th}}$. Suppose that g and h are known, that g'' exists throughout $\mathscr{I}$, and that $g' > 0$ on $\mathscr{I}$. From (9.2) and (9.20) we see that *in $\mathscr{D}_{\text{th}}$ the pressure function ϖ determines Λ_V uniquely and determines K_V to within a function of temperature alone.* Therefore, if g, h, and $\mathscr{D}_{\text{th}}$ satisfy the conditions just stated and belong in common to two bodies whose pressure functions differ by a function of V alone, then the latent heats Λ_V of these bodies are the same, while their specific heats K_V differ by a function of temperature alone. We leave it to the reader to verify that the assumption $g' > 0$ is not essential to the truth of this conclusion.

Remark 2. For the caloric theory of LAPLACE and CARNOT, use of (6.4) in the second equation of (9.13) shows that $\phi = 0$, and so the first equation of (9.16) shows that $h = 1$. In fact, the conditions (9.1) and $h = 1$ are valid not merely in a thermodynamic part $\mathscr{D}_{\text{th}}$ but throughout $\mathscr{D}$. When specialized to the caloric theory, (9.2) reduces to the celebrated *Carnot–Clapeyron Theorem*:

$$\frac{\partial \varpi}{\partial \theta} = g' \Lambda_V. \tag{9.21}$$

This relation played a central part in the early studies of thermodynamics, though its logical status has remained obscure. In some of the early literature it was interpreted as asserting no more than the existence of a function of temperature, say C, such that

$$\frac{\partial \varpi}{\partial \theta} = C \Lambda_V. \tag{9.22}$$

This relation, in which C need not be g', may be considered on its own merits, with no commitment for or against the caloric theory. Theorem 7 asserts that in a thermodynamic part (9.22) holds always, and that $C = g'/h$.

Historical Comment. It is the function C, whether or not $C = g'$, that KELVIN and JOULE called "Carnot's function".

Corollary 7.5. *Along an adiabat in $\mathscr{D}_{\text{th}}$*

$$\frac{g'}{h}\frac{d\theta}{dV} = -\frac{1}{K_V}\frac{\partial \varpi}{\partial \theta}. \tag{9.23}$$

Along an adiabat in a part of $\mathscr{D}_{\text{th}}$ where γ is of one sign

$$\frac{g'}{h}\frac{d\theta}{dp} = -\frac{1}{K_p}\frac{\partial \varpi/\partial \theta}{\partial \varpi/\partial V}. \tag{9.24}$$

Proof. To obtain (9.23), we multiply (4.3) by g'/h and then use (9.2). The same procedure applied to (4.7), followed by use of the first equation of (3.9), yields (9.24). $\quad\square$

Corollary 7.6. *Let* $\theta = f(V)$ *represent an adiabat in* $\mathcal{D}_{\text{th}}$. *Then at a piezotropic point on the adiabat*

$$g' \frac{d^2\theta}{dV^2} = -\frac{h}{K_V} \frac{\partial^2 \varpi}{\partial V \partial \theta}. \tag{9.25}$$

If $g' \neq 0$ *at the piezotropic point, this expression delivers the curvature of the adiabat at the piezotropic point. If* g *is twice continuously differentiable on* $\mathcal{I}$, *the parametrization of a piezotrope in* $\mathcal{D}_{\text{th}}$ *satisfies the differential relation* (2.8).

> *Proof.* To obtain (9.25), we differentiate (4.3) and then use (9.2). The last sentence is immediate from the existence and continuity of $\partial^2 \varpi / \partial V \partial \theta$ and $\partial^2 \varpi / \partial \theta^2$ in $\mathcal{D}_{\text{th}}$. □

Remark. These two corollaries, unlike the statements about adiabats and piezotropes in Chapters 2 and 4, rest upon thermodynamics, not merely the Doctrine of Latent and Specific Heats. The former expresses what seem to be the only thermodynamic aspects of the "anomalous behavior" of water that may be found in the standard books on thermodynamics.[2] At points where $g' \neq 0$ of course (9.25) is merely another expression for (4.4).

Theorem 8 (Pro-Entropy and Internal Pro-Energy). *On* $\mathcal{D}_{\text{th}}$ *there is a pro-entropy* H_h *such that*

$$\Lambda_V = h \frac{\partial H_h}{\partial V}, \qquad K_V = h \frac{\partial H_h}{\partial \theta}; \tag{9.26}$$

also, there is an internal pro-energy $E_{g,h}$ *such that*

$$\frac{g}{h} \Lambda_V = \varpi + \frac{\partial E_{g,h}}{\partial V}, \qquad \frac{g}{h} K_V = \frac{\partial E_{g,h}}{\partial \theta}. \tag{9.27}$$

Both H_h *and* $E_{g,h}$ *are twice continuously differentiable. For a process in* $\mathcal{D}_{\text{th}}$

$$Q = h\dot{H}_h, \qquad \dot{E}_{g,h} = \frac{g}{h} Q - \varpi \dot{V} \tag{9.28}$$

almost always. In particular, a process in $\mathcal{D}_{\text{th}}$ *is adiabatic if and only if* $H_h = \text{const.}$ *for it.*

> *Proof.* Theorem 7 asserts that the condition (9.1) is satisfied at every point of $\mathcal{D}_{\text{th}}$. Since $\mathcal{D}_{\text{th}}$ is simply connected, this condition implies the existence of a function $H_h(V, \theta)$ on $\mathcal{D}_{\text{th}}$ satisfying (9.26).

---

2. References to the thermodynamics of anomalous behavior are rare in the textbooks. An exception is that of J. KESTIN, cited in Footnote 9 to Chapter 2. Of course, since he deals only with CLAUSIUS' thermodynamics, for purposes of comparison we have to set $g'/h = J/\theta$, J being a positive constant (*cf.* Chapter 15). In his §12.7, through some unnecessary manipulations of partial derivatives, KESTIN obtains as his Equation (12.46a) a result equivalent to (9.24) in a part of $\mathcal{D}_{\text{th}}$ where $\partial \varpi / \partial \theta$ is of one sign, thus tacitly excluding the piezotropic points. Of course, by a somewhat different chain of identities he could have obtained from his starting point a result fully equivalent to the case of (9.24) that is appropriate to classical thermodynamics.

The first equation of (9.18) may be written in the form

$$\frac{\partial}{\partial \theta}\left(\frac{g}{h}\Lambda_V - \varpi\right) = \frac{\partial}{\partial V}\left(\frac{g}{h}K_V\right). \tag{9.29}$$

This condition implies that on $\mathscr{D}_{\text{th}}$ a function $E_{g,h}(V, \theta)$ such as to satisfy (9.27) exists. The relations (9.28) follow immediately from (3.1), (9.26), and (9.27). Because ϖ, Λ_V, and K_V have been assumed continuously differentiable in $\mathscr{D}$, while g and h have been proved to be continuously differentiable on $\mathscr{I}$, (9.26) and (9.27) show that the partial derivatives of H_h and $E_{g,h}$ have continuous derivatives in $\mathscr{D}_{\text{th}}$. $\square$

Remark 1 (Entropy, Heat Function).   From the first equation of (9.28) we see that *h is an integrating factor for the heating* in all processes which traverse a set of points in $\mathscr{D}_{\text{th}}$. Thus $H_h(V, \theta)$ is a corresponding pro-entropy in $\mathscr{D}_{\text{th}}$. In Chapter 6 we noted the local existence of an integrating factor $f(V, \theta)$ in general, but we have made no use of that fact. We have proved directly that for all processes in $\mathscr{D}_{\text{th}}$ the function h, which is a function of θ alone, is an integrating factor for Q. In the remarks after Definition 13 in Chapter 6 we noted that the choices $h = 1$ and $h = \theta$ are appropriate for the caloric theory and classical thermodynamics, respectively. The corresponding functions H_1 and H_θ are both special cases of the pro-entropy H_h, cases that follow from the choices of the function h permitted by the two theories, which are known to contradict each other. In this sense the *entropy H_θ* of CLAUSIUS is a replacement for the *heat function H_1* of CARNOT. For this reason and others TRUESDELL has often referred to the entropy by the more suggestive word *calory*.

Remark 2 (Internal Energy).   The internal energy of classical thermodynamics is a special case of $E_{g,h}$ for an appropriate choice of the functions g and h (*cf.* Corollary 15.1 of Chapter 15). In the caloric theory the internal pro-energy is $E_{g,1}$. By use of (9.26) and (9.27) we can show that $E_{g,1}$ is related to the heat function as follows:

$$\frac{\partial E_{g,1}}{\partial V} = g\frac{\partial H_1}{\partial V} - \varpi,$$

$$\frac{\partial E_{g,1}}{\partial \theta} = g\frac{\partial H_1}{\partial \theta}. \tag{9.30}$$

These relations hold also in the general theory if $E_{g,1}$ and H_1 are replaced by $E_{g,h}$ and H_h, respectively.

Historical Comment.   In 1850 CLAUSIUS obtained (9.27) in the special case of an ideal gas, on the assumption that heat and work were interconvertible by all cyclic processes. In 1854 he obtained (9.27) more generally along with results he much later put into the form (9.26), using of course the particular functions g and h which are appropriate to classical thermodynamics. Meanwhile REECH in 1853 had published both (9.26) and (9.27) in the more general but incompletely specified form that results from use of his function Γ. Neither of those authors obtained the full results we give here. CLAUSIUS, for his part, showed no sign of seeing that in the caloric theory there was an internal pro-energy $E_{g,1}$; REECH did not perceive that Q had an integrating factor which was a function of temperature alone. Since much of the contents of Theorem 8 may be inferred by combining and comparing the results of CLAUSIUS and REECH, we may with some justice name it the *Clausius–Reech Theorem* or REECH's *Third Theorem*.

Remark 3 (Interconvertibility of Heat and Work). From the second equation of (9.28) it is clear that in the theory based on CARNOT's Axiom, the heating contributes to the growth of $E_{g,h}$ only through the term $(g/h)Q$, the factor g/h being a function of temperature alone. According to classical thermodynamics we are allowed to choose g and h so that $g/h = J$, a positive constant, the same for all bodies. Then the contribution of a given heating Q to the growth of the internal energy so selected is independent of the temperature. This assumption, which expresses the *universal as well as uniform interconvertibility of heat and work*, we shall take up in Chapter 10 and again in Chapter 15.

Remark 4 (Totality of Pro-Entropies and Internal Pro-Energies). The notations for the pro-entropy and internal pro-energy indicate the dependence on the choice of the functions g and h. Even when g and h have been selected, the pro-entropy and the internal pro-energy are determined only to within additive constants. The functions g and h whose existence was demonstrated in Theorem 7 are not unique. If the continuously differentiable functions $\bar{g}$ and $\bar{h}$ replace g and h, respectively, in (9.1) and (9.2), from Theorem 7 we know that $\bar{g}$ and $\bar{h}$ are related to g and h by (9.17). An easy calculation based on (9.26) and (9.27) shows that the pro-entropy $H_{\bar{h}}$ and the internal pro-energy $E_{\bar{g},\bar{h}}$ corresponding to $\bar{g}$ and $\bar{h}$ are related to H_h and $E_{g,h}$ by

$$H_{\bar{h}} = \frac{H_h}{K} + \text{const.,}$$

$$E_{\bar{g},\bar{h}} = E_{g,h} + \frac{N}{K} H_h + \text{const.,}$$

(9.31)

K and N being the constants in (9.17).

We introduce the notation

$$\Delta f \equiv f(t_2) - f(t_1).$$

(9.32)

From the second equation of (9.31) we notice that

$$\Delta E_{\bar{g},\bar{h}} - \Delta E_{g,h} = (N/K)\Delta H_h = N\Delta H_{\bar{h}}.$$

(9.33)

Integrating the second equation of (9.28) from t_1 to t_2, we obtain:

Corollary 8.1. *For a process in $\mathcal{D}_{\text{th}}$*

$$\Delta E_{g,h} = \int_{t_1}^{t_2} \left(\frac{g}{h} - J\right) Q dt + JC - L, \qquad (9.34)$$

J being an arbitrary constant. In particular, for an adiabatic process

$$L = -\Delta E_{g,h}; \qquad (9.35)$$

that is, in an adiabatic process in $\mathcal{D}_{\text{th}}$ a body does work at the expense of its internal pro-energy.

Remark. A body undergoing a process in which the heating Q and the working $\varpi \dot{V}$ are both almost always zero is said to be *isolated*. For an isolated body, from the second equation of (9.28) we see that $\dot{E}_{g,h} = 0$. That is, the *internal pro-energy of an isolated body remains constant*. This fact is some-times[3] called "the law of conservation of energy". A still different idea is the uniform interconvertibility of heat and work, mentioned in Remark 3 after Theorem 8: There is a function E, defined over $\mathcal{D}$, such that in any process

$$\Delta E = JC - L, \qquad (9.36)$$

J being a positive constant. If $\mathcal{D}$ is a thermodynamic part, Corollary 8.1 asserts that (9.36) holds for all adiabatic processes. If in addition $g = Jh + \text{const.}$, then Corollary 8.1 and the first equation of (9.28) show that (9.36) holds for all processes.

Corollary 8.2 (Condition Imposed by the Pressure Function on the Internal Pro-Energy).

$$g\frac{\partial \varpi}{\partial \theta} - g'\varpi = g'\frac{\partial E_{g,h}}{\partial V}. \qquad (9.37)$$

Proof. To obtain (9.37), we multiply the first equation of (9.27) by g' and use (9.2). □

Corollary 8.3. *In a thermodynamic part $\mathcal{D}_{\text{th}}$ which is isothermally and isochorically convex,[4] the following four conditions are equivalent:*

$$K_V = k(\theta). \qquad (9.38)$$

$$\Lambda_V = h(\theta)l(V). \qquad (9.39)$$

---

3. For example, by M. PLANCK in §66 of his *Treatise on Thermodynamics*, transl. A. OGG, 3rd ed., Longmans, London, 1926, although in §58 PLANCK gives the name "princi-ple of the conservation of energy" to a much vaguer statement, which presupposes the uniform and universal interconvertibility of heat and work.
4. A subset of $\mathcal{D}$ is *isochorically convex* if whenever it contains two distinct points on the same isochor it contains also the segment of the isochor that joins them. Compare Footnote 1 to this chapter. A four-pointed star whose axes are isothermal and isochoric segments is both isochorically and isothermally convex but is not convex.

$$E_{g,h} = f(V) + \int \frac{k(\theta)g(\theta)}{h(\theta)} \, d\theta. \tag{9.40}$$

$$\varpi = -f'(V) + g(\theta)l(V). \tag{9.41}$$

Proof. The result follows by use of (9.1), (9.2), and (9.27). ☐

Remark. Let (9.38)–(9.41) be satisfied. From (4.1), (9.38), and (9.39) we conclude that a process in $\mathcal{D}_{\text{th}}$ is adiabatic if and only if

$$\int l(V)dV + \int \frac{k(\theta)}{h(\theta)} \, d\theta = \text{const.} \tag{9.42}$$

From (9.39) and (9.41) it is clear that l and f' are continuously differentiable functions. Use of (9.38), (9.39), and (9.41) in the second equation of (3.9) then yields

$$\gamma - 1 = \frac{[g'(\theta)h(\theta)/k(\theta)][l(V)]^2}{f''(V) - g(\theta)l'(V)}. \tag{9.43}$$

The thermodynamic structure derived in this chapter lies at two levels: first, the local relations connecting ϖ, Λ_V, K_V, g, and h, and, second, the properties of the internal pro-energy $E_{g,h}$ and the pro-entropy H_h. It seems that a thermodynamic part $\mathcal{D}_{\text{th}}$ is nearly the most general subset of $\mathcal{D}$ on which relations of both kinds can be shown to subsist as a consequence of Axioms I, II, and III.

The constitutive domain $\mathcal{D}$ may have several thermodynamic parts, none of which is a subset of the others. For example, a multiply connected $\mathcal{D}$ cannot be a thermodynamic part, but it may well be the union of two or more overlapping thermodynamic parts. As these subsets can be chosen in infinitely many different ways, it should be possible to find uniform local relations over $\mathcal{D}$. More generally, the thermodynamic parts of $\mathcal{D}$ may make up a set that is not even connected; they may be scattered over $\mathcal{D}$ like spots. Nevertheless, it is possible to demonstrate a *local thermodynamics* valid throughout all of them. That is what we shall now do.

Definition 17$_{\text{bis}}$. A thermodynamic point is *normal* if the isotherm passing through it contains an ordinary point of $\mathcal{D}$. The set of all normal thermodynamic points of $\mathcal{D}$ is the *normal set* $\mathcal{D}_n$ of $\mathcal{D}$. A simple cyclic process that traverses and includes only points of $\mathcal{D}_n$ is a *simple cyclic process in $\mathcal{D}_n$*.

Every ordinary point is a normal thermodynamic point. All the neutral points of the constitutive domain shown in Figure 7A in Chapter 5 are thermodynamic points, but none of them is a normal thermodynamic point. On the other hand, every neutral point of the constitutive domains in Figures 7B and 7C is also a normal thermodynamic point.

$\mathcal{D}_n$ need not be open; it may be empty. A thermodynamic part $\mathcal{D}_{\text{th}}$ of $\mathcal{D}$ as defined by Definition 17 is a nonempty, simply connected, open subset of $\mathcal{D}_n$. The set of points included by a simple cyclic process in $\mathcal{D}_n$ is a simply connected open subset of $\mathcal{D}_n$.

Theorem 7 delivers constitutive restrictions that hold at every point of a thermodynamic part of $\mathcal{D}$. A small modification of the proof of that theorem enables us to extend these restrictions to every *normal thermodynamic point* of $\mathcal{D}$.

Let $\mathcal{O}$ denote the collection of temperatures corresponding to the ordinary points of $\mathcal{D}$. Since the set of ordinary points is open, $\mathcal{O}$ is an open set, possibly empty, of positive real numbers. Thus $\mathcal{O}$ is the union of a countable collection[5] of disjoint open intervals $\mathcal{I}_k$. The intervals $\mathcal{I}_k$ need not be bounded. In the proof of Theorem 7 we constructed functions f and ϕ defined on the interval of temperatures corresponding to $\mathcal{D}_{\text{th}}$. The Third Principal Lemma enables us to extend these functions to the set $\mathcal{O}$ as follows. If θ_0 is any temperature in $\mathcal{O}$, the isotherm $\theta = \theta_0$ contains at least one ordinary point, and we define $f(\theta_0)$ as the common value of $(\partial\varpi/\partial\theta)/\Lambda_V$ at all ordinary points of $\mathcal{D}$ on the isotherm $\theta = \theta_0$; $\phi(\theta_0)$, as the common value of $(\partial\Lambda_V/\partial\theta - \partial K_V/\partial V)/\Lambda_V$ at all ordinary points on the isotherm $\theta = \theta_0$. Any temperature θ in $\mathcal{O}$ lies in one and only one of the intervals $\mathcal{I}_k$. We select arbitrarily some one temperature from each $\mathcal{I}_k$ and so define on $\mathcal{O}$ a function that has a constant value $\theta_k(\theta)$ on each $\mathcal{I}_k$. The method used in the proof of Theorem 7 enables us to extend to $\mathcal{O}$ the functions g and h:

$$h(\theta) \equiv \exp \int_{\theta_k(\theta)}^{\theta} \phi(u)\,du,$$

$$g(\theta) \equiv \int_{\theta_k(\theta)}^{\theta} h(u)f(u)\,du,$$

(9.44)

$$k = 1, 2, 3, \ldots.$$

It is a routine matter to extend to these more general functions g and h the arguments we used to prove Theorem 7. First we show that the extended functions f and ϕ are continuous on $\mathcal{O}$ and that $f > 0$ except on a subset of $\mathcal{O}$ having empty interior. The definitions (9.44) then make h a continuously differentiable positive function and g a continuously differentiable function which increases on each interval $\mathcal{I}_k$, though not necessarily on $\mathcal{O}$. Since the extended functions f and ϕ satisfy (9.15) at every ordinary point of $\mathcal{D}$, also the basic constitutive restrictions (9.1) and (9.2) hold at every ordinary point provided that the functions g and h appearing in them be interpreted as the extensions defined by (9.44). The relations (9.1) and (9.2) subsist by continuity at every normal thermodynamic point since the temperature corresponding to any such point belongs to $\mathcal{O}$.

5. See, for example, pp. 37–38 of B. Sz.-NAGY, *Introduction to Real Functions and Orthogonal Expansions*, New York, Oxford University Press, 1965.
Of all the theorems of mathematics we apply in this tractate this is the only one that was not at the disposal of the pioneers of thermodynamics. However, there is nothing in it to surprise any mathematically sensitive person, then or now. As in some other aspects of explicit rigor, the difference is more one of style than of content.
The teacher of an elementary course today need not make much of the matter. He may easily render the statement of the theorem plausible by examples. If he prefers, he may avoid the point altogether by resting content to restrict attention to the most general kind of domain that has so far occurred in applications: one for which $\mathcal{O}$ is the union of a finite number of $\mathcal{I}_k$.

Suppose that a positive differentiable function $\bar{h}$ and a differentiable function $\bar{g}$ on $\mathcal{O}$ be such as to replace g and h in (9.1) and (9.2) at every ordinary point. Then we easily show that they are related to g and h by (9.17) in each interval $\mathcal{I}_k$, with possibly different positive constant K and constant N for each interval $\mathcal{I}_k$. Conversely, any two functions $\bar{g}$ and $\bar{h}$ on $\mathcal{O}$ such as to satisfy (9.17) in each interval $\mathcal{I}_k$ for some constant N and some positive constant K, these constants not necessarily being the same for each of the intervals, may replace g and h in (9.1) and (9.2) at every ordinary point.

The foregoing arguments prove the following extension of Theorem 7:

Theorem 7$_{\text{ext}}$ (Basic Constitutive Restrictions in $\mathcal{D}_n$). *Let $\mathcal{O}$ be the set of temperatures corresponding to the ordinary points of $\mathcal{D}$; let $\mathcal{I}_k$ denote any member of the countable collection of disjoint open intervals into which $\mathcal{O}$ may be decomposed. On $\mathcal{O}$ there are continuously differentiable functions g and h, the former being an increasing function on each $\mathcal{I}_k$ and the latter a positive one, such that at every point of $\mathcal{D}_n$*

$$\frac{\partial}{\partial \theta} \left(\frac{\Lambda_V}{h} \right) - \frac{\partial}{\partial V} \left(\frac{K_V}{h} \right) = 0, \tag{9.1}_r$$

$$\frac{g'}{h} \Lambda_V = \frac{\partial \varpi}{\partial \theta}. \tag{9.2}_r$$

In each interval $\mathcal{I}_k$ the function h is unique to within a constant factor; when h has been determined, g is unique in $\mathcal{I}_k$ to within an additive constant.

The most important consequence of Theorem 7$_{\text{ext}}$ is that we may select a *fixed* pair of functions g and h such as to render (9.1) and (9.2) valid in *every* thermodynamic part and also on and within *every* ordinary Carnot cycle. This result is beyond the reach of the Second Principal Lemma and Theorem 7. Making use of Theorem 7$_{\text{ext}}$, we easily compile a list of extended results.

1. In the statements of Corollaries 7.1, 7.2, and 7.5, replace $\mathcal{D}_{\text{th}}$ by $\mathcal{D}_n$ and replace $\mathcal{I}$ by $\mathcal{O}$.
2. In the statements of Corollaries 7.3, 7.4, and 7.6, replace $\mathcal{D}_{\text{th}}$ by any nonempty open subset of $\mathcal{D}_n$, and replace $\mathcal{I}$ by the collection of temperatures corresponding to that open set.
3. In the caloric theory the Carnot–Clapeyron Theorem (9.21) holds at every point of $\mathcal{D}_n$.

By use of (9.2) as extended we infer a simple property of the commonest kind of constitutive domain, exemplified by ideal gases and Van der Waals fluids:

4. If $\mathcal{D}$ has no piezotropic points, then $g' > 0$ on $\mathcal{O}$, and a *thermodynamic point is normal if and only if it is ordinary*. For such a $\mathcal{D}$ the set $\mathcal{D}_n$ is open, and a neutral thermodynamic point cannot be normal.

CHAPTER 10

Cyclic Processes in $\mathcal{D}_{\text{th}}$ and $\mathcal{D}_{\text{n}}$.
C-Processes.
Efficiency Theorem and Emission–
Absorption Theorem for C-Processes.
Completeness Theorems.

In Chapter 8 the First Principal Lemma and its Corollary refer only to cycles in a Carnot web. Now, after three preliminary lemmas, we shall introduce a class of cyclic processes called C-processes; this class includes Carnot processes as a special case. We shall then extend the central results of Chapter 8 to C-processes in $\mathcal{D}_{\text{th}}$; our aim is to prove two major theorems: the Emission–Absorption Theorem and the Efficiency Theorem.

Lemma 1. *Let $\mathcal{C}$ be a simple cycle in $\mathcal{D}$ such that the constitutive restrictions (9.1) and (9.2) hold at all points lying on it or included by it. For any process belonging to $\mathcal{C}$*

$$L = \int_{t_1}^{t_2} \frac{g}{h} Q\, dt, \qquad \int_{t_1}^{t_2} \frac{Q}{h}\, dt = 0. \tag{10.1}$$

In particular, *these relations hold for all simple cyclic processes in $\mathcal{D}_{\text{n}}$.*

> *Proof.* The first displayed equation follows by applying AMPÈRE's transformation to the first equation of (9.18); the second follows similarly from (9.1). □

Lemma 2. *The relations (10.1) hold for any cyclic process in $\mathcal{D}_{\text{th}}$, and so does the following:*

$$L = \int_{t_1}^{t_2} g\dot{H}_h dt. \tag{10.2}$$

> *Proof.* These relations follow from (9.28). □

Remark 1. The reader will note that we have demonstrated (10.1) from two different assumptions. The second proof does not require that the cyclic process be simple, but it must traverse only *interior points* of $\mathscr{D}_n$; the first proof requires the process to be *simple* but permits it to traverse boundary points of $\mathscr{D}_n$.

While a similar distinction may be made in much of what follows now, we choose to present the complete theory only for processes in $\mathscr{D}_{th}$. At the end of this chapter we shall list the verbal substitutions sufficient to convert it to simple cyclic processes in $\mathscr{D}_n$.

Remark 2. From the first equation of (10.1) it is clear that the work done by a body undergoing a cyclic process in $\mathscr{D}_{th}$ is determined by the latent and specific heats of that body, provided the function g/h be known. The pro-entropy H_h and the function g also determine the work done, as is shown by (10.2).

Remark 3. Since we have shown that h is an integrating factor for the heating in $\mathscr{D}_{th}$, all the results derived in Chapter 6 are proved valid in $\mathscr{D}_{th}$ if in them we replace f by h. For example, we could read off the second equation in (10.1) as a special case of (6.12). We shall now restate Theorem 5 in this way, noting that since h is a function of θ alone, the symbols "max", "min", "sup", "inf" may be regarded as referring to subsets of $\mathscr{I}$.

Lemma 3. *For a cyclic process in $\mathscr{D}_{th}$*

$$\frac{\max h}{\min h} C^+ \geqq C^- \geqq \frac{\min h}{\max h} C^+. \qquad (10.3)$$

Equality subsists on the right-hand side if and only if $h = \max h$ on $\mathscr{T}^+$ and $h = \min h$ on $\mathscr{T}^-$; on the left-hand side if and only if $h = \min h$ on $\mathscr{T}^+$ and $h = \max h$ on $\mathscr{T}^-$. If the sets $\mathscr{T}^+$ and $\mathscr{T}^-$ are not empty,

$$\frac{\sup_{\mathscr{T}^-} h}{\inf_{\mathscr{T}^+} h} C^+ \geqq C^- \geqq \frac{\inf_{\mathscr{T}^-} h}{\sup_{\mathscr{T}^+} h} C^+. \qquad (10.4)$$

Equality subsists on either side if and only if $h = $ const. on $\mathscr{T}^+$ and on $\mathscr{T}^-$.

Remark. The properties of cyclic processes mentioned in Remark 1 after Theorem 5 in Chapter 6 are valid for every cyclic process in $\mathscr{D}_{th}$. Namely, cyclic processes are of *two kinds only*:

1. *Adiabatic* cyclic processes. From (10.2) we see that $L = 0$ for these.
2. Cyclic processes that *both absorb and also emit positive amounts of heat*: $C^+ > 0$, $C^- > 0$. For these, L may be positive, negative, or null.

In particular, *a body cannot do work in a cyclic process unless it not only absorbs a positive amount of heat but also emits a positive amount.*

This last is one of the most celebrated conclusions of classical thermo-dynamics. As our proof of it shows, it is common to all theories based on CARNOT's General Axiom.

Definition 18. A cyclic process for which neither $\mathscr{T}^+$ nor $\mathscr{T}^-$ is empty is a *C-process* if it absorbs heat at one and only one temperature θ^+ and emits heat at one and only one temperature θ^-. That is,

$$\begin{aligned} \theta &= \theta^+ \text{ on } \mathscr{T}^+, \\ \theta &= \theta^- \text{ on } \mathscr{T}^-. \end{aligned} \tag{10.5}$$

The temperature θ^+ is the *absorption temperature*; θ^-, the *emission tempera-ture*.

Remark. For a C-process, as for a Carnot process, both the heat absorbed C^+ and the heat emitted C^- are positive. Comparing Definition 18 with Definition 14 in Chapter 7, we see that a Carnot process is a C-process in which $\theta^- < \theta^+$. It is immediate from the definition that the reverse of a C-process is also a C-process. In particular, the reverse of a Carnot process is a C-process, although, as already noted in Chapter 7, it is not a Carnot process. For a C-process θ^+ need not differ from θ^-. Figure 10A in Chapter 7 illustrates a C-process in which $\theta^- = \theta^+$.

Theorem 9 (Emission–Absorption Theorem for C-Processes in $\mathscr{D}_{\text{th}}$). *For a C-process in* $\mathscr{D}_{\text{th}}$

$$C^- = \frac{h(\theta^-)}{h(\theta^+)} C^+. \tag{10.6}$$

Proof. For a C-process h is constant on $\mathscr{T}^+$ and on $\mathscr{T}^-$, and (10.6) follows at once from the second relation in (10.1). $\square$

Remark. The role of h in (10.6) suggests that we call it the *heat-loss function*.

Corollary 9.1 (Absorption–Emission Inequality). *The following three conditions are equivalent in* $\mathscr{D}_{\text{th}}$:

1. *For every ordinary Carnot cycle* $C^+ \geqq C^-$.
2. *The function* h *is nondecreasing.*
3. *For every Carnot process* $C^+ \geqq C^-$.

Proof. First we show that Condition 1 implies Condition 2. Let B be a temperature in $\mathscr{I}$. The isotherm $\theta = B$ contains an ordinary point of $\mathscr{D}_{\text{th}}$. Therefore, we can construct a Carnot web $\mathscr{W}_{A,B}$ generated by an ordinary Carnot cycle in $\mathscr{D}_{\text{th}}$ if A is a temperature in $\mathscr{I}$ such that $B - A$ is a sufficiently small positive number. Because Condition 1 holds for the cycles of the web $\mathscr{W}_{A,B}$, it follows from (10.6) that h is nondecreasing on $[A, B]$. Hence $h'(B) \geqq 0$. Since B is an arbitrary temperature in $\mathscr{I}$, $h' \geq 0$ throughout $\mathscr{I}$. Therefore, h is nondecreasing on $\mathscr{I}$. That Condition 2 implies Condition 3, is an immediate consequence of (10.6). Clearly Condition 3 implies Condition 1. $\square$

Corollary 9.2. *The following three conditions are equivalent in $\mathscr{D}_{th}$:*

1. *For every ordinary Carnot cycle $C^+ > C^-$.*
2. *The function h is increasing.*
3. *For every Carnot process $C^+ > C^-$.*

Proof. We begin by showing that Condition 1 implies Condition 2. By the preceding corollary, $h' \geq 0$ throughout $\mathscr{I}$. If h' vanishes on a subinterval contained in $\mathscr{I}$, we can construct a Carnot web whose cycles $\mathscr{C}$ satisfy $C^+(\mathscr{C}) = C^-(\mathscr{C})$, violating Condition 1. Hence $h' > 0$ except on a subset of $\mathscr{I}$ with empty interior. Condition 2 now follows. The rest of the proof is similar to the preceding one. $\square$

Remark. If h is an increasing or decreasing function, the condition $h = $ const. on a subset of $[t_1, t_2]$ is equivalent to $\theta = $ const. on that subset. In that case we can conclude that equality results on either side of (10.4), and hence on both sides, for C-processes and for them alone.

Corollary 9.3 (Characterization of the Caloric Theory). *The following three conditions are equivalent in $\mathscr{D}_{th}$:*

1. *For every ordinary Carnot cycle $C^+ = C^-$.*
2. *The function h is constant.*
3. *For every cyclic process $C^+ = C^-$.*

Proof. That Condition 1 implies Condition 2, follows by reasoning similar to that used in the proof of Corollary 9.1. To show that Condition 2 implies Condition 3, we use the second equation of (10.1). Clearly Condition 3 implies Condition 1. $\square$

Theorem 10 (Efficiency Theorem for C-Processes in $\mathscr{D}_{th}$). *For a C-process in $\mathscr{D}_{th}$*

$$L = \frac{g(\theta^+) - g(\theta^-)}{h(\theta^+)} C^+. \tag{10.7}$$

Proof. We can write the first equation of (10.1) in the form

$$L = \int_{\mathscr{I}^+} \frac{g}{h} Q dt - \int_{\mathscr{I}^-} \frac{g}{h}(-Q)dt. \tag{10.8}$$

For a C-process, by definition

$$\frac{g}{h} Q = \frac{g(\theta^+)}{h(\theta^+)} Q \text{ on } \mathscr{I}^+, \quad \frac{g}{h}(-Q) = \frac{g(\theta^-)}{h(\theta^-)}(-Q) \text{ on } \mathscr{I}^-. \tag{10.9}$$

Substitution of (10.9) into (10.8) yields

$$L = \frac{g(\theta^+)}{h(\theta^+)} C^+ - \frac{g(\theta^-)}{h(\theta^-)} C^-. \tag{10.10}$$

Elimination of C^- from (10.10) and (10.6) leads to (10.7). $\square$

Corollary 10.1. *For all arguments* x, y, z *that can correspond to an ordinary Carnot cycle in* $\mathcal{D}_{th}$, *the Carnot function G in Axiom III is given by*

$$G(x, y, z) = \frac{g(x) - g(y)}{h(x)} z. \tag{10.11}$$

Remark 1. Indeed, combining the results of Theorem 6 and the First Principal Lemma, we get a result that has the same form as (10.7) or (10.11) and holds for all cycles in a Carnot web, but there the functions g and h were shown to exist only over the interval $[A, B]$ corresponding to the web; even at that, it remained possible that two different webs $\mathscr{W}^1_{A,B}$ and $\mathscr{W}^2_{A,B}$ might lead to different functions g and h. The functions g and h in Theorem 10 are defined over the entire temperature interval $\mathscr{I}$ of $\mathcal{D}_{th}$, and (10.7) applies to all *C*-processes in $\mathcal{D}_{th}$. The result (10.10) serves to extend (8.17) to *C*-processes in $\mathcal{D}_{th}$.

Remark 2. From Definition 18 it follows immediately that a process equivalent to a *C*-process is also a *C*-process having the same absorption and emission temperatures. If we call the cycle generated by a *C*-process a *C*-cycle, from the reversal theorems it is clear that the results (10.6) and (10.7), established above for *C*-processes in $\mathcal{D}_{th}$, are valid also for *C*-cycles in $\mathcal{D}_{th}$.

Remark 3. Since g is an increasing function, from (10.7) we conclude that the sign of the work done by a body which undergoes a *C*-process in $\mathcal{D}_{th}$ is the sign of $\theta^+ - \theta^-$. Thus *a body in undergoing a Carnot process in* $\mathcal{D}_{th}$ *does positive work*. This result extends the inequality in (8.1), which was postulated for ordinary Carnot cycles alone. A trivial corollary of (10.6) and (10.7) deserves some notice[1]: If in a *C*-process $\theta^+ = \theta^-$, then $C^+ = C^-$, and $L = 0$. A process belonging to the simple cycle shown in Figure 10A of Chapter 7 is clearly a *C*-process in which $\theta^+ = \theta^-$.

Remark 4. The results (10.6) and (10.7) hold for all *C*-processes in $\mathcal{D}_{th}$ regardless of whether they be simple or not. These results hold also for all types of Carnot cycles in $\mathcal{D}_{th}$, including those shown as Examples (C), (D), and (E) in Figure 9 of Chapter 7, which are not ordinary.

Corollary 10.2 (Position of CARNOT's Special Axiom, REECH). *The following conditions are equivalent in* $\mathcal{D}_{th}$:

1. *For every cyclic process* $C^+ = C^-$.

1. The supertrivial conclusion that $C^+ = C^-$ and $L = 0$ in an isothermal cycle was called "Moutier's Theorem" by J. R. PARTINGTON in §II30 of *An Advanced Treatise on Physical Chemistry*, Volume 1, London *etc.*, Longmans, Green & Co., 1949; he cited a work of J. MOUTIER published in 1875, far too late to belong to the origins of thermodynamics. We are not sure what PARTINGTON meant when he wrote, "The theorem does not follow from the *First* Law, as erroneously stated...," since little if anything follows from any *single* axiom. PARTINGTON did not make clear the assumptions from which he derived his conclusion. Because it does hold in CARNOT's theory, which contradicts the First Law, it plainly does not *require* that the First Law hold.

2. *There is a function $\bar{g}$ on $\mathscr{I}$ such that*

$$G(x, y, z) = [\bar{g}(x) - \bar{g}(y)]z \qquad (10.12)$$

for all arguments x, y, z that correspond to some ordinary Carnot cycle.

> *Proof.* Condition 1 implies that h is constant on $\mathscr{I}$ by Corollary 9.3. That Condition 1 implies Condition 2, then follows from Corollary 10.1. We shall now show that Condition 2 implies Condition 1. We construct a Carnot web $\mathscr{W}_{A,B}$ as in the proof of Corollary 9.1. Applying (10.12) and (10.11) to the cycles of this web, we show that if $B \geqq x \geqq y \geqq A$
>
> $$\bar{g}(x) - \bar{g}(y) = \frac{g(x) - g(y)}{h(x)}. \qquad (10.13)$$
>
> This is the special case of (8.10) that corresponds to $\bar{h} = 1$. From Remark 1 after the First Principal Lemma in Chapter 8 we see that $\bar{g}$ is continuously differentiable and h is constant on $[A, B]$. Hence $h' = 0$ throughout $\mathscr{I}$; that is, h is constant on $\mathscr{I}$. Application of Corollary 9.3 completes the proof. $\square$

Remark 1.   The caloric theory took it as an axiom[2] that $C^+ = C^-$ for every cyclic process. In CARNOT's words,[3]

> ...after a complete circle of operations the gas is brought back identically to its original condition. It has had to give up all the caloric that was furnished to it at the outset.

CARNOT himself, although at one place he stated Axiom III in words, in all his calculations used only the special case (10.12) in place of the equation in (8.1), and at that only for infinitesimally small values of $x - y$, and of course he assumed in effect that all of $\mathscr{D}$ was a thermodynamic part. The preceding corollary shows that (10.12) is *a necessary and sufficient condition for Axioms* I, II, *and* III *to be compatible with the caloric theory of heat*, if $\mathscr{D}$ is assumed to be a thermodynamic part.

Historical Scholion.   Because most of CARNOT's reasoning rests upon (2.9), (3.1), and Axiom III with G of the form (10.12), not appealing directly to the indestructibility of heat, many students of thermodynamics have thought that most of his results were independent of any such assumption. Thus physicists often claim that CARNOT "knew" the Second Law without "knowing" the First, and so "obtained correct results from an incorrect axiom". Corollary 10.2 shows that such is not the case. CARNOT had neither

2. In the remark following Definition 13 in Chapter 6 the caloric theory was characterized by a different axiom, but the two axioms can be shown to be equivalent.
3. *Réflexions*, p. 57. Compare also p. 37:
We assume implicitly in our demonstration that when a body has suffered any changes whatever and after a certain number of transformations is brought back to its initial state, that is, to that state considered relatively to the density, the temperature, the manner of aggregation—we assume, I say, that the body will be found to contain the same quantity of heat as it did at first, or, in other words, that the quantities of heat absorbed or emitted in its various transformations are exactly compensated.

the "First Law" nor the "Second Law". His working assumptions (8.1) and (10.12) *are not compatible with* Axioms I and II unless $C^+ = C^-$ *in every cyclic process.* We have reached this conclusion essentially through the use of Theorem 10, which neither CARNOT nor REECH possessed. Our treatment here, as at many other points, could have been much simpler had we been content, as CARNOT and REECH were, to obtain local results alone and to leave unspecified any conditions such as to ensure that ordinary Carnot cycles exist in sufficient abundance. CARNOT's results, except those few that are independent of the choice of g and h, *cannot be freed* from the assumption, which he did indeed lay down explicitly, that heat is conserved. Apart from a few of his statements of inequality discussed below, we have found only four claims of CARNOT that do not require for their truth the caloric theory of heat. Of course, one of these is his General Axiom (8.1); the others are the Theorem on Progressions, quoted below, just after Property 6 in Chapter 11, the result quoted in the Historical Comment after Property 4 in Chapter 11, and the result given in the Historical Comment after Theorem 13 in Chapter 13.

As we have stated in Remark 1, CARNOT was careful to limit his own specific applications of his ideas to Carnot cycles whose operating temperatures were nearly equal. For them, obviously, the distinction between his General Axiom and his Special Axiom is blurred. Consequently, since CLAUSIUS' theory conforms with the General Axiom, the distinction between CLAUSIUS' theory and CARNOT's is likewise blurred. Near any one temperature, both will yield just the same results to within a factor of proportionality. Only by comparing the factors appropriate to different ranges of temperature can different results emerge. This fact seems to have escaped the notice of historians of science, some of whom have made much of the status of CARNOT's theory as an approximation to CLAUSIUS', regarding it as the result of a sort of divination.

Remark 2. Define the function $F(x, y)$ for x and y in $\mathcal{I}$ as follows:

$$F(x, y) \equiv \frac{g(x) - g(y)}{h(x)}. \tag{10.14}$$

In terms of this function (10.7) can be written as

$$L = F(\theta^+, \theta^-)C^+. \tag{10.15}$$

From (8.4) and (8.8) it is clear that the function defined by (10.14) is an extension of the function F introduced in Theorem 6 of Chapter 8. It is clear that Theorems 10 and 9 extend the results in the First Principal Lemma and its Corollary in Chapter 8 to C-processes in $\mathcal{D}_{th}$. Carnot processes and ordinary Carnot cycles in $\mathcal{D}_{th}$ are included as special cases. If x and y are selected arbitrarily from $\mathcal{I}$, we have no assurance that there is a C-process in $\mathcal{D}_{th}$ having x as its absorption temperature and y as its emission temperature. If such a process exists, then for it (10.6) and (10.7) hold.

Remark 3. CARNOT seems to have assumed $\mathscr{D}$ to be such that ordinary Carnot cycles having arbitrary operating temperatures chosen from $\mathscr{I}$ always exist. If we grant that such be the case, we may use (10.15) to interpret certain claims made by CARNOT[4]:

> The motive power doubtless increases with the difference of temperatures of the hot body and the cold body, but we do not know if it be proportional to this difference.

One interpretation for the first part of this statement would be: $F(x, y)$ is an increasing function of x when $x > y$ and y is fixed, and a decreasing function of y when $y < x$ and x is fixed. Since g is an increasing function, the latter of these assertions is confirmed by (10.14), but the former does not generally follow from it. However, this interpretation of CARNOT's claim does hold in his own special case of the theory, for which h is constant; it holds also whenever the function G is invariant under change of the unit of temperature (compare Theorem 16, Chapter 16).

Another statement of CARNOT, one on which he laid emphasis, is[5]

> The fall of caloric produces more motive power in the lower ranges than in the higher ones.

That is, $F(x + k, x)$ is a decreasing function of x for any fixed, positive k less than the length of the interval $\mathscr{I}$. This conclusion does not generally follow from (10.14) even for the caloric theory; however, it does follow when G is restricted to functions invariant under change of the unit of temperature (compare Theorem 16, Chapter 16).

Corollary 10.3 (Characterization of Uniform Interconvertibility of Heat and Work). *Let J be a positive constant. The following conditions are equivalent in* $\mathscr{D}_{\text{th}}$:

1. *If $\mathscr{C}$ is an ordinary Carnot cycle, then $L(\mathscr{C}) = JC(\mathscr{C})$.*
2. *On the temperature interval $\mathscr{I}$*

$$g = Jh + \text{const.} \tag{10.16}$$

3. *For every cyclic process $L = JC$.*

> *Proof.* We begin by proving that Condition 1 implies Condition 2. Constructing a Carnot web $\mathscr{W}_{A,B}$ as in the proof of Corollary 9.1 and applying (10.7), (10.6), and Condition 1 to the cycles of this web, we show that if $B \geqq x \geqq y \geqq A$,
>
> $$\frac{g(x) - g(y)}{h(x)} = J\left[1 - \frac{h(y)}{h(x)}\right]. \tag{10.17}$$
>
> This is a special case of (8.10) corresponding to the choice $\bar{g} = Jh$, $\bar{h} = h$. From the first equation of (8.12) and (8.14) it is clear that $g = Jh + \text{const.}$ on $[A, B]$. Hence $g' = Jh'$ throughout $\mathscr{I}$, and

4. *Réflexions*, pp. 28–29.
5. *Réflexions*, p. 72.

Condition 2 follows. To show that Condition 2 implies Condition 3, we substitute (10.16) in the first equation of (10.1) and use the second equation. Clearly Condition 3 implies Condition 1. □

Remark 1. Condition 3 of Corollary 10.3 expresses the *uniform interconvertibility of heat and work* in cyclic processes, J being the *mechanical equivalent of a unit of heat*. In the remark after Corollary 8.1 in Chapter 9 uniform interconvertibility of heat and work was characterized by the existence of a function E such as to satisfy (9.36) in every process. It is easy to see that this characterization is equivalent to Condition 3 of Corollary 10.3.

We have stated in the remark following Lemma 3 at the beginning of this chapter that if $L > 0$ for a cyclic process in $\mathscr{D}_{th}$, then $C^- > 0$ and $C^+ > 0$. Since $C = C^+ - C^-$, from Condition 3 we read off another celebrated statement of classical thermodynamics: *In a cyclic process, a body that interconverts heat and work uniformly can never do as much work as the mechanical equivalent of the heat it absorbs.* Thus any heat engine labors under an inherent disadvantage.

Remark 2. If $C^+ > 0$, we may express Condition 3 of Corollary 10.3 as follows:

$$\frac{L}{JC^+} = 1 - \frac{C^-}{C^+}. \tag{10.18}$$

In some works the efficiency of a body undergoing a cyclic process is defined by the right-hand side of this relation.

Remark 3. Corollary 10.3 establishes (10.16) as a necessary and sufficient condition that heat and work be uniformly interconvertible by cyclic processes in $\mathscr{D}_{th}$. Through the use of (9.2), (9.29), and (9.1) we obtain the following two conditions, each of which is equivalent to (10.16):

$$\frac{\partial \varpi}{\partial \theta} = J \frac{h'}{h} \Lambda_V,$$

$$\frac{\partial}{\partial \theta} (\varpi - J\Lambda_V) = -J \frac{\partial K_V}{\partial V}. \tag{10.19}$$

The second of these, which is due to CLAUSIUS, is universal in the sense that it is free of the function h. A typical use of it is as follows. For a certain body, let $\mathscr{D}$ have a thermodynamic part $\mathscr{D}_{th}$ that is isothermally and isochorically convex. Then *in order for K_V to be a function of θ alone in $\mathscr{D}_{th}$ it is necessary and sufficient that $\varpi - J\Lambda_V$ be a function of V alone there.*

Remark 4. Since g is an increasing function, a glance at (10.16) shows that for a body which interconverts heat and work uniformly h also must be an increasing function. For such a body the function F in (10.14) is given by

$$F(x, y) = J\left[1 - \frac{h(y)}{h(x)}\right]. \tag{10.20}$$

Remark 5. Consider a body whose constitutive domain $\mathscr{D}$ contains an ordinary point. Then $\mathscr{D}$ contains a thermodynamic part $\mathscr{D}_{th}$. If this body interconverts heat and work uniformly, h is an increasing function on the temperature interval of $\mathscr{D}_{th}$. If, on the other hand, the body obeys the caloric theory of heat, then $h = $ const. For such a body, therefore, the two principal theories of heat cannot both be true. Thus the first equation of (10.19) and the Carnot–Clapeyron Theorem (9.21) cannot both hold. Despite their formal similarity, they are mutually exclusive.

Scholion. Our analysis refers to a particular fluid body. To say that that body interconverts work and heat *uniformly* means that its action in this regard is independent of the temperatures at which it operates. J. R. MAYER (1842), on the basis of philosophic speculation, and JOULE (1847), on the basis of careful and systematic experiments in many particular cases, asserted that *all bodies had this property*. That is, heat and work are interconvertible not only uniformly but also *universally*. This hypothesis was adopted as an axiom by RANKINE (1850), CLAUSIUS (1850), and KELVIN (1851), and classical thermodynamics, which was constructed in one way or another by those three savants, rests upon it. In the context of the present theory, it requires that in all cyclic processes $L = JC$ with *one and the same positive constant J* for all bodies. Hence it requires that for every body the functions g and h corresponding to a thermodynamic part determine each other mutually through (10.16). We do not adopt that axiom in this tractate. Rather, we shall prove it as a part of Corollary 15.1 in Chapter 15.

Corollary 10.4 (Characterization of Classical Efficiency). *Let J be a positive constant. The following conditions are equivalent in $\mathscr{D}_{th}$:*

1. *For every ordinary Carnot cycle $\mathscr{C}$ having θ^+ and θ^- as its operating temperatures*

$$L(\mathscr{C}) = J\left(1 - \frac{\theta^-}{\theta^+}\right)C^+(\mathscr{C}). \tag{10.21}$$

2. *On the interval $\mathscr{I}$*

$$g = Jh + \text{const.}, \qquad h = M\theta, \qquad M = \text{const.} > 0. \tag{10.22}$$

3. *For every C-process having θ^+ and θ^- as its absorption and emission temperatures, respectively,*

$$L = J\left(1 - \frac{\theta^-}{\theta^+}\right)C^+. \tag{10.23}$$

Proof. To show that Condition 1 implies Condition 2, we begin by constructing a Carnot web $\mathscr{W}_{A,B}$ as in the proof of Corollary 9.1. Applying (10.7) and Condition 1 to the elements of $\mathscr{W}_{A,B}$, we show that if $B \geqq x \geqq y \geqq A$,

$$\frac{g(x) - g(y)}{h(x)} = J\left(1 - \frac{y}{x}\right). \tag{10.24}$$

This is a special case of (8.10) corresponding to the choice $\bar{g}(x) = Jx$ and $\bar{h}(x) = x$. From the first equation of (8.12) and (8.14) it follows that $g' = Jh'$ and $h(x) = Mx$ on $[A, B]$, M being a positive constant. By a similar argument applied to a Carnot web $\mathscr{W}_{B,C}$, C being a temperature in $\mathscr{I}$ sufficiently close to B, we conclude that $h' = \text{const.}$ on $[B, C]$. Since h' is continuous, it is constant on $[A, C]$. In particular, h' is differentiable at B, and $h''(B) = 0$. Since B is an arbitrary temperature in $\mathscr{I}$, $h'' = 0$ throughout $\mathscr{I}$; that is, there are constants a and b such that $h(x) = ax + b$ in $\mathscr{I}$. Comparing this expression for h in $\mathscr{I}$ with the expression $h(x) = Mx$ on $[A, B]$ obtained earlier, we conclude that $a = M$ and $b = 0$; that is, $h(x) = Mx$ for all x in $\mathscr{I}$. Since $g' = Jh'$ throughout $\mathscr{I}$, the first equation of (10.22) follows. By substituting (10.22) into (10.7) we show that Condition 2 implies Condition 3. Clearly Condition 3 implies Condition 1. $\square$

Corollary 10.5. *A body cannot satisfy the classical efficiency condition* (10.21) *in all ordinary Carnot cycles unless it interconverts heat and work uniformly while undergoing processes in any thermodynamic part. If the condition* (10.21) *is* universal, *that is, common to all bodies in ordinary Carnot cycles, then all bodies interconvert heat and work uniformly while undergoing processes in a thermodynamic part.*

Proof. The result follows by comparing the second condition of Corollary 10.4 with the second condition of Corollary 10.3. $\square$

Corollary 10.6. *If a body satisfies the condition* (10.21) *in all ordinary Carnot cycles, then for a C-process in* $\mathscr{D}_{\text{th}}$

$$\frac{C^-}{C^+} = \frac{\theta^-}{\theta^+},\tag{10.25}$$

θ^+ *and* θ^- *being the absorption and emission temperatures, respectively.*

Proof. We have only to substitute the second equation of (10.22) into (10.6). $\square$

Scholion. The efficiency corresponding to (10.21), namely $1 - \theta^-/\theta^+$, is that nowadays commonly accepted for "reversible" Carnot cycles. It was first obtained by KELVIN (1851, imperfectly), RANKINE (1851), and CLAUSIUS (1856), all of whom laid down as an axiom the universal and uniform interconvertibility of heat and work. The course of our argument, on the contrary, cleaves to the earlier tradition, which sought to found thermodynamics upon the properties of heat engines. In Chapter 15 we shall see that the universal choice (10.22) for the functions g and h implies the entire classical thermodynamics of "reversible" processes in fluid bodies. Thus Corollary 10.4 achieves the aim that CARNOT, CLAPEYRON, and REECH sought but failed to attain. To those whose bent is more practical than speculative, the argument given above may seem superior to CLAUSIUS', since it is based from first to last, not on general assertions about nature, but rather upon the motive power of Carnot cycles, nothing more.

Although the outright assumption that $1 - \theta^-/\theta^+$ is the universal efficiency of Carnot cycles affords the shortest path to classical thermodynamics for those who are already convinced that that theory is what they wish, we prefer to obtain the same ultimate result through a deeper analysis of the general theory based on CARNOT's General Axiom.

Before proceeding to impose our final axioms we assert and prove a theorem that shows we have exhausted the content of CARNOT's General Axiom, as far as a thermodynamic part of $\mathscr{D}$ is concerned, in two ways: (1) by the local conditions upon the constitutive functions ϖ, Λ_V, and K_V; and (2) by expressing the constitutive functions in terms of the pro-entropy and the internal pro-energy.

Theorem 11 (Completeness of the Constitutive Restrictions and the Energy–Entropy Relations in $\mathscr{D}_{\mathrm{th}}$). *Let Axioms I and II be assumed. Then in a thermodynamic part $\mathscr{D}_{\mathrm{th}}$, the temperature interval of which is $\mathscr{I}$, the following three conditions are equivalent:*

1. CARNOT's *General Axiom (our Axiom III).*
2. *On $\mathscr{I}$ there are continuously differentiable functions g and h such that* $h(x) > 0$, $(x - y)(g(x) - g(y)) > 0$ *if $x \neq y$, and*

$$\frac{\partial}{\partial \theta}\left(\frac{\Lambda_V}{h}\right) - \frac{\partial}{\partial V}\left(\frac{K_V}{h}\right) = 0, \tag{9.1}_\mathrm{r}$$

$$\frac{g'}{h}\Lambda_V = \frac{\partial \varpi}{\partial \theta}. \tag{9.2}_\mathrm{r}$$

3. *On $\mathscr{I}$ there are continuously differentiable functions g and h such that* $h(x) > 0$, $(x - y)(g(x) - g(y)) > 0$ *if $x \neq y$, and on $\mathscr{D}_{\mathrm{th}}$ there are functions H and E such that everywhere*

$$\Lambda_V = h\frac{\partial H}{\partial V}, \qquad K_V = h\frac{\partial H}{\partial \theta}, \tag{10.26}$$

$$\frac{g}{h}\Lambda_V = \varpi + \frac{\partial E}{\partial V}, \qquad \frac{g}{h}K_V = \frac{\partial E}{\partial \theta}. \tag{10.27}$$

Proof. Theorem 7 in Chapter 9 asserts that Condition 1 implies Condition 2. Because the proof of Theorem 8 in Chapter 9 is based entirely on (9.1) and (9.2), it is clear that Condition 2 implies Condition 3. Now suppose Condition 3 is satisfied. Then for any process in $\mathscr{D}_{\mathrm{th}}$ we may substitute (10.26) and (10.27) into (3.1) and so obtain almost always

$$Q = h\dot{H}, \qquad \frac{g}{h}Q = \varpi\dot{V} + \dot{E}. \tag{10.28}$$

Therefore for a cyclic process in $\mathscr{D}_{\mathrm{th}}$

$$L = \int_{t_1}^{t_2}\frac{g}{h}Qdt, \qquad \int_{t_1}^{t_2}\frac{Q}{h}dt = 0, \tag{10.1}_\mathrm{r}$$

the first being a consequence of (1.6). Applying (10.1) to a C-process, we obtain (10.6) and (10.10); combining these, we obtain (10.7). By assumption g is an increasing function, so (10.7) shows that $L > 0$ for any Carnot process. In particular, then, for any ordinary Carnot cycle $\mathscr{C}$ it follows that

$$L(\mathscr{C}) = \frac{g(\theta^+) - g(\theta^-)}{h(\theta^+)} C^+(\mathscr{C}) > 0, \qquad (10.29)$$

so CARNOT's General Axiom (8.1) is satisfied (and the function G which occurs in it is shown to be of a particular kind). $\square$

Remark. If $\mathscr{D}$ is a thermodynamic part, by replacing $\mathscr{D}_{\text{th}}$ by $\mathscr{D}$ in Theorem 11 we see that the constitutive restrictions expressed by Condition 2 or Condition 3 of the theorem *exhaust the content of* CARNOT's *General Axiom.* A slight modification of Condition 3 also leads to a sufficient condition for Axiom III without placing any restrictions on the constitutive domain $\mathscr{D}$, as shown by the following corollary.

Corollary 11.1 (Sufficiency of the Energy–Entropy Formulation even if $\mathscr{D}$ is not a Thermodynamic Part). *Let Axioms* I *and* II *be assumed. If on the temperature interval of $\mathscr{D}$ there are continuously differentiable functions g and h, g being an increasing function and h a positive one, while on $\mathscr{D}$ there are functions H and E such that (10.26) and (10.27) hold at every point of $\mathscr{D}$, then* CARNOT's *General Axiom (our Axiom* III*) is satisfied with the function G given by (10.11).*

    Proof. Referring to the proof of Theorem 11, by an argument parallel to that used there in proving that Condition 3 implies Condition 1 we show that (10.7) holds for any C-process. The required result then follows. $\square$

Remark. A somewhat different line of argument yields a more general result, as follows.

Corollary 11.2 (Sufficiency of the Constitutive Restrictions even if $\mathscr{D}$ is not a Thermodynamic Part). *Let Axioms* I *and* II *be assumed. If on the temperature interval of $\mathscr{D}$ there are continuously differentiable functions g and h, g being an increasing function and h a positive one, such as to satisfy the constitutive restrictions (9.1) and (9.2) at every ordinary point of $\mathscr{D}$, then* CARNOT's *General Axiom is satisfied with the function G given by (10.11).*

    Proof. By Lemma 1 the relations (10.1) hold for every simple cyclic process that includes thermodynamic points alone. It is then easy to verify that (10.6) and (10.7) hold for every simple C-cycle whose interior consists of thermodynamic points alone. Therefore, (10.29) holds for every ordinary Carnot cycle $\mathscr{C}$. $\square$

Remark 1. In the modern approaches to classical thermodynamics it is usual to make in effect a special and universal choice of the functions g and h, namely, $g = Jh$ and $h = \theta$, and then to postulate axioms sufficient to ensure that every body has an internal energy E and an entropy H such as to make (10.26) and (10.27) valid at every point of a constitutive domain. Corollary 11.1 shows that if Axioms I and II are presumed to be valid, such a theory is always compatible with CARNOT's General Axiom, the function G being given by $G(x, y, z) = J(1 - (y/x))z$ for all bodies. However, the sufficient condition in the corollary is not necessary for Axiom III, as is shown by the following example of a fluid body: $\mathscr{D}$ = entire V-θ quadrant, $\varpi = \theta/V$, $\Lambda_V = 0$, and $K_V = V$. This example is compatible with Axioms I and II and also, since no Carnot cycles for it exist, with Axiom III. If (10.26) holds for this example, then (9.1) also holds, leading to the contradiction $1/h = 0$. This example shows also that the statement of Theorem 11 becomes false in general if $\mathscr{D}_{th}$ is replaced by $\mathscr{D}$ and if $\mathscr{I}$ is interpreted as the temperature interval of $\mathscr{D}$.

Remark 2. Corollary 11.2 enables us to construct examples of fluid bodies compatible with Axioms I, II, and III that contradict certain claims made by CARNOT, provided, at least, our earlier interpretations of those claims be just. We begin with the following example: $\mathscr{D}$ = entire V-θ quadrant, $\varpi = \theta + 1/V$, $\Lambda_V = \theta^2$, and $K_V = 1$. By choosing $g = \theta$ and $h = \theta^2$ it is easy to verify that the condition in Corollary 11.2 is satisfied. Hence this example is compatible with Axiom III. It is of course compatible with Axioms I and II. Since the adiabats are given by $V = 1/\theta + $ const., CARNOT's assumption that ordinary Carnot cycles exist for any pair of operating temperatures is also satisfied. For this example, the function F defined by (10.14) has the form $F(x, y) = (x - y)/x^2$ and is not an increasing function of x when $x > y$ and y is fixed. We have therefore exhibited a fluid body that is compatible with Axioms I, II, and III yet contradicts the first statement of CARNOT quoted in Remark 3 after Corollary 10.2. The second statement quoted there is also generally false even for the caloric theory, as is shown by the example: $\mathscr{D}$ = entire V-θ quadrant, $\varpi = \theta + 1/V$, $\Lambda_V = 1$, and $K_V = 1$. For this example the condition in Corollary 11.2 is satisfied by choosing $g = \theta$ and $h = 1$. Therefore this example is compatible with Axioms I, II, and III and also with the caloric theory of heat. The adiabats are given by $V + \theta = $ const., and hence ordinary Carnot cycles having arbitrary operating temperatures exist. However, $F(x + k, x) = k$, and hence $F(x + k, x)$ is not a decreasing function of x for any fixed k. This example shows also that the function g might be linear in the caloric theory, as CARNOT himself remarked (*Réflexions*, pp. 70 and 79), and in that case the motive power of a given quantity of heat for a given "fall of caloric" would be the same in all ranges of temperature.

Remark 3. Let Axioms I and II be assumed. Then by the Second Principal Lemma of Chapter 9, Axiom III implies that (9.1) and (9.2) are satisfied at every ordinary point of $\mathscr{D}$, g and h being locally defined functions depending on the ordinary point. Corollary 11.2 may be regarded as a converse of this result when g and h exist throughout the temperature interval of $\mathscr{D}$. A body for which $\Lambda_V = 0$ at every point of $\mathscr{D}$ trivially satisfies Axiom III, because the set of Carnot cycles is then empty. We note that Corollary 11.2 covers this degenerate case.

Remark 4. The arguments in Remark 2 after Theorem 6 in Chapter 8, when applied to (10.15), show that $F(x, y)/J$ is the efficiency of a body undergoing in $\mathscr{D}_{th}$ a C-process whose absorption temperature is x and emission temperature is y. For the example given just before Remark 3, $F(x, y)/J = (x - y)/J$. For this example all simple Carnot cycles are ordinary, Carnot cycles exist for any pair of operating temperatures, and the efficiency of a Carnot cycle tends to infinity as $x \to \infty$ while y remains fixed. Thus CARNOT's theory does not impose any limitation upon the efficiency of engines. This fact does not contradict any of the statements CARNOT published in his treatise. Indeed,[6] according to CARNOT,

> Thus the production of motive power in steam engines is due, not to a real consumption of caloric, but *to its transport from a hot body to a cold one...*

However, we must not conclude that CARNOT's theory necessarily yields large efficiencies. Consider the example: $\mathscr{D}$ = entire $V\text{-}\theta$ quadrant, $\varpi = 1 - \exp(-\theta) + 1/V$, $\Lambda_V = 1$, and $K_V = 1$. For this example the condition in Corollary 11.2 is satisfied by choosing $g = 1 - \exp(-\theta)$ and $h = 1$. Thus this example is compatible with Axioms I, II, and III and also with the caloric theory. It is also easy to verify that every simple Carnot cycle is ordinary and that Carnot cycles having arbitrary operating temperatures exist. However, the efficiency of a Carnot cycle is given by $[\exp(-y) - \exp(-x)]/J$ and hence is never so great as $1/J$.

The proper choice of $\bar{g}$ in (10.12) occasioned CARNOT and his successors great concern.

Remark 5. In the ordinary writings on thermodynamics we may read occasional appeals to the "absurdity" of something or other. For example, it would be "absurd" if an engine could give out both heat and work at the same time. One such absurdity is excluded by Axioms I and II, for Corollary 2.1 of Chapter 3 and (2.15) show that if a sequence of ordinary Carnot cycles $\mathscr{C}$ belonging to a Carnot web is such that $C^+(\mathscr{C}) \to 0$, then $L(\mathscr{C}) \to 0$ for the sequence. However, CARNOT's General Axiom is not strong enough to forbid the amount of heat absorbed in a Carnot cycle from being less than that emitted.[7] Therefore, CARNOT's General Axiom does not require that Condition 1 of Corollary 9.1 hold. Of course, we may impose this condition as an additional axiom if we wish to exclude the particular "absurdity" in question. There are also those who regard the caloric theory of heat as absurd. If $\mathscr{D}$ is a thermodynamic part, they may exclude it by imposing Condition 1 of

---

6. *Réflexions*, p. 10.
7. The example: $\mathscr{D}$ = entire $V\text{-}\theta$ quadrant, $\varpi = \theta + 1/V$, $\Lambda_V = 1/\theta$, and $K_V = 1$ satisfies the condition in Corollary 11.2 if we choose $g = \theta$ and $h = 1/\theta$. Hence the example is compatible with all the axioms. However, by (10.6) we see that $C^+(\mathscr{C}) < C^-(\mathscr{C})$ for every Carnot cycle $\mathscr{C}$.

Corollary 9.2 as an axiom; for them, *both g and h are increasing functions of θ*. In fact, however, no rejection of "absurdities" is necessary to obtain all the classical results, for, as we shall see in Chapter 15, they are mathematical consequences of two further simple and natural axioms, both of them affirmative.

Remark 6. The function g appearing in Theorem 7 of Chapter 9 is increasing on $\mathscr{I}$, and therefore $g' > 0$ except on a subset of $\mathscr{I}$ with empty interior. However, we may not generally exclude the possibility that $g'(\theta) = 0$ for certain particular values of θ. To make this clear we construct another example: $\mathscr{D} =$ entire V-θ quadrant, $\varpi = (\theta - 1)^3 + 1/V$, $\Lambda_V = 1$, and $K_V = 1$. For this example the condition in Corollary 11.2 is satisfied if we choose $g = (\theta - 1)^3$ and $h = 1$. Thus this example, for which $g'(1) = 0$, is compatible with all the axioms and even with the caloric theory. It shows also that Λ_V need not vanish at every piezotropic point, because the isotherm $\theta = 1$ consists of piezotropic points that are not neutral points (*cf.* Corollary 7.2 in Chapter 9).

Corollary 11.3 (Sufficiency of the Determination of Λ_V and K_V from ϖ, g, and h). *Let Axioms I and II be assumed. Let $\mathscr{I}$ be the temperature interval of $\mathscr{D}$. If g is a twice differentiable increasing function on $\mathscr{I}$, if h is a continuously differentiable positive function on $\mathscr{I}$, and if (9.2) and (9.19) are satisfied at every point of $\mathscr{D}$, then* CARNOT's *General Axiom (our Axiom III) holds.*

> *Proof.* Assume first that $g' > 0$. Then we can solve for Λ_V/h and $\partial K_V/\partial V$ from (9.2) and (9.19), respectively, and verify that (9.1) is satisfied. Since every point where $g' = 0$ is the limit of a sequence of points at which $g' > 0$, it follows by continuity that (9.1) subsists throughout $\mathscr{D}$. The result then follows by noting that the condition in Corollary 11.2 is satisfied. $\square$

Extension of Corollary 8.3 to Piezotropic Fluids. A body of piezotropic fluid is one for which the constitutive domain $\mathscr{D}$ is a piezotropic part. Lemma 1 of Chapter 8 shows that for such a fluid, $\mathscr{D}$ is also a neutral part, that is, $\Lambda_V = 0$ in $\mathscr{D}$. This much we can derive from CARNOT's General Axiom; because there are no Carnot cycles for such a fluid, we can derive nothing more. In order to get a thermodynamics sufficient to lay further restrictions upon piezotropic fluids, we must abandon any axiomatic structure based on Carnot cycles. One way to do so is simply to *assume* as an axiom that with each fluid body are associated continuously differentiable functions g and h on the temperature interval $\mathscr{I}$ of $\mathscr{D}$, g being an increasing function and h a positive one, and functions H and E on $\mathscr{D}$ such as to satisfy (10.26) and (10.27) in $\mathscr{D}$. If we take this axiom in place of Axiom III, Corollary 11.1 shows that the statement of Axiom III follows for a special choice of G. Therefore, all the results obtained in this tractate on the basis of Axioms I, II, and III follow also from Axioms I and II, and the newly stated axiom. Thus the new theory is compatible with the one we are constructing in this tractate. Further, it is easy to verify that if $E_{g,h}$ is replaced by E in (9.40) the four conditions in Corollary 8.3 continue to be equivalent for all bodies for which $\mathscr{D}$ is isothermally and isochorically convex. For a piezotropic fluid in particular, (9.39) is always true, because we can choose $l = 0$. If the constitutive domain of the piezotropic fluid is also isothermally and isochorically convex, the extended Corollary 8.3 enables us to conclude that

$$K_V = k(\theta),$$

$$E = f(V) + \int \frac{k(\theta)g(\theta)}{h(\theta)}\, d\theta, \tag{10.30}$$

$$\varpi = -f'(V),$$

for the fluid. Of course for such a fluid $\gamma = 1$.

Although the theory so obtained does make it unnecessary to restrict attention almost entirely to thermodynamic parts of $\mathscr{D}$, it is a patchwork of two different lines of thought and hence pleonastic. If we are willing to introduce energy and entropy straight off, we can produce a cleaner and far more general theory almost effortlessly. That theory we shall present in Chapter 17 as the epilogue to this tractate.

At the end of Chapter 9 we introduced the normal set $\mathscr{D}_n$ and extended the local relations (9.1) and (9.2) to it. Lemmas 1 and 2 at the beginning of this chapter indicate two different possibilities for deriving properties of cyclic processes, one for simple cyclic processes in $\mathscr{D}_n$ and the other for cyclic processes in $\mathscr{D}_{th}$. There we chose to exploit only the latter possibility. Now we fulfil our promise to provide verbal substitutions such as to convert results established for $\mathscr{D}_{th}$ into results valid for $\mathscr{D}_n$.

The most important example is provided by Theorems 9 and 10. Both of these were proved by appeal to (10.1) only. For (10.1), regarded as a pair of formulae, we have given two different sufficient conditions, which we have stated in Lemma 1 and Lemma 2, respectively. Theorems 9 and 10 followed from Lemma 2 alone. Turning back now to Lemma 1, we read off

Theorems 9_{bis} and 10_{bis} (Emission–Absorption and Efficiency Theorems in $\mathscr{D}_n$). *For a simple C-process in $\mathscr{D}_n$*

$$C^- = \frac{h(\theta^-)}{h(\theta^+)}\, C^+, \tag{10.6}_r$$

$$L = \frac{g(\theta^+) - g(\theta^-)}{h(\theta^+)}\, C^+. \tag{10.7}_r$$

Corollary 10.1_{ext}. *The Carnot function G in Axiom* III *is given by*

$$G(x, y, z) = \frac{g(x) - g(y)}{h(x)}\, z \tag{10.11}_r$$

for all arguments x, y, z in its domain of definition.

Thus the formula that appears in the First Principal Lemma in Chapter 8 and in Corollary 10.1 is extended to *all arguments for which the function G is defined*. The pair of functions g and h, which are defined on all of $\mathcal{O}$, may be selected once and for all. Our freedom in choice of a particular pair has been delimited in the argument leading to Theorem 7_{ext}.

We leave it to the reader to verify that Lemma 3, Corollaries 9.1–9.3, and Corollaries 10.2–10.6 subsist if in their statements we make the following replacements:

1. Replace $\mathcal{D}_{\text{th}}$ by $\mathcal{D}_{\text{n}}$.
2. Replace "cyclic process" by "simple cyclic process".
3. Replace "nondecreasing" by "nondecreasing on each interval $\mathcal{S}_k$".
4. Replace "increasing" by "increasing on each interval $\mathcal{S}_k$".
5. Replace "constant" by "constant on each $\mathcal{S}_k$", the only exception to this rule being the constant J appearing in Corollaries 10.3 and 10.4.
6. Replace "the interval $\mathcal{S}$" by "the set $\mathcal{O}$".

With the apparatus we have constructed it is easy to show that *if the constitutive restrictions* (9.1) *and* (9.2) *hold at all ordinary points of* $\mathcal{D}$, *then they exhaust the content of* CARNOT's *General Axiom*. Indeed, if we lay down only Axioms I and II and assume that there are continuously differentiable functions g and h, g being an increasing function on each $\mathcal{S}_k$ and h being a positive function, such as to satisfy (9.1) and (9.2) at every ordinary point of $\mathcal{D}$, it follows from Lemma 1 that (10.1) holds for every simple cyclic process in $\mathcal{D}_n$. Thus (10.6) and (10.7) hold for every ordinary Carnot cycle in $\mathcal{D}$. Furthermore, the work done by any ordinary Carnot cycle is *positive*, for g, by assumption, increases on each $\mathcal{S}_k$. We have thus shown that Axiom III is satisfied with G given by (10.11). By combining this result with Theorem 7_{ext} we arrive at the following *necessary and sufficient* condition for CARNOT's General Axiom to hold when Axioms I and II are assumed:

Corollary 11.2$_{\text{ext}}$ (Equivalence of the Constitutive Restrictions and CARNOT's General Axiom). *Let Axioms I and II be assumed. Let $\mathcal{S}_k$ denote any interval in the decomposition of the open set $\mathcal{O}$ of temperatures corresponding to the ordinary points of $\mathcal{D}$. Then* CARNOT's *General Axiom (our Axiom III) is satisfied if and only if there are continuously differentiable functions g and h on $\mathcal{O}$, g being an increasing function on each interval $\mathcal{S}_k$ and h a positive one, such as to satisfy (9.1) and (9.2) at every ordinary point of $\mathcal{D}$; then the function G in Axiom III is given by (10.11).*

We remark that the restrictions of the constitutive functions ϖ, Λ_V, and K_V to the set of all ordinary points of a given body compatible with Axioms I and II determine whether or not that body be also compatible with Axiom III. This result is to be expected since a neutral point cannot lie on or within any ordinary Carnot cycle, while CARNOT's General Axiom refers to ordinary Carnot cycles alone. Of course we do not mean to say that CARNOT's General Axiom fails to restrict ϖ, Λ_V, and K_V at neutral points. Indeed, we have shown that the axiom implies (9.2) at all normal thermodynamic points, so a neutral normal thermodynamic point is necessarily a piezotropic point.

So much for the local relations and their consequences. It is clear that the results concerning the pro-entropy and the internal pro-energy cannot be extended, in general, to $\mathscr{D}_n$.

CHAPTER 11

Properties of Ideal Gases and Van der Waals Fluids.

Ideal Gases

The ideal gas has always played a central role in the foundations of classical thermodynamics. Two important properties of ideal gases have been demonstrated and recorded in Chapter 3; another, at the end of Chapter 7. Ever since CARNOT's treatise was published, it has been clear that the specific heats of an ideal gas, which are connected through thermodynamic relations but not determined by them, open the door to the general theory. Although we shall defer our consideration of this fact until Chapters 15 and 16, here we derive ten more *properties of ideal gases*.

We recall that an ideal gas is defined as being a body having the thermal equation of state (2.9). In this chapter we assume that the constitutive domain $\mathscr{D}$ of any body considered is a thermodynamic part. In Theorem 11 we have seen that for such a body the constitutive restrictions (9.1) and (9.2) are necessary and sufficient for CARNOT's General Axiom to be satisfied. Here we shall study the consequences of these restrictions of thermodynamics when applied to ideal gases and Van der Waals fluids.

Property 5. For a body of ideal gas the function g' is positive and continuously differentiable;

$$\Lambda_V = \frac{Rh}{Vg'} > 0; \tag{11.1}$$

$$\frac{\partial K_V}{\partial V} = -\frac{Rhg''}{Vg'^2}; \tag{11.2}$$

$$K_p - K_V = \frac{Rh}{\theta g'} > 0; \tag{11.3}$$

and if $\mathcal{D}$ is isothermally convex,

$$K_V = -\frac{Rhg''}{g'^2} \log V + K, \tag{11.4}$$

K being a function of θ alone.

> *Proof.* Substitution of (2.9) into (9.2) yields $g'\Lambda_V = Rh/V$. We know that $g' \geq 0$. Therefore both g' and Λ_V are positive functions, and (11.1) follows. Since $g' = (Rh)/(V\Lambda_V)$, it is clear that g' is continuously differentiable. The other results follow by use of (2.9) in (9.19), the second equation of (3.9), and (9.20). $\square$

Remark 1. From Property 5 it is clear that for an ideal gas every point of $\mathcal{D}$ is an ordinary point, $\gamma > 1$, and $K_p - K_V$ is a function of temperature alone.

Remark 2. It does not follow that if one specific heat of an ideal gas is constant, also the other must be. Indeed, in CARNOT's theory, for which we can choose $h = 1$, it is clear from (11.2) and (11.3) that if $K_V = $ const., then K_p cannot be constant.

Property 6 (CARNOT's Theorem on the Isothermal Absorption of Heat by an Ideal Gas). The heat absorbed by a body of ideal gas on a simple path $\mathcal{P}_\theta$ of isothermal processes from the volume V_a to the greater volume V_b at the temperature θ is given by

$$C^+(\mathcal{P}_\theta) = \frac{h(\theta)}{g'(\theta)} R \log \frac{V_b}{V_a}. \tag{11.5}$$

> *Proof.* We need only substitute (11.1) into (3.7). $\square$

Remark. According to CARNOT,[1]

> When a gas changes in volume without change of temperature, the quantities of heat absorbed or emitted by that gas are in arithmetic progression if the increments or decrements of volume are found to be in geometric progression.

Property 6 affirms that CARNOT's conclusion remains valid not only in the caloric theory but also in any thermodynamics based on CARNOT's General Axiom.

Property 7. For a body of ideal gas the following conditions are equivalent:

1. K_V is a function of θ alone.
2. K_p is a function of θ alone.
3. γ is a function of θ alone.

> *Proof.* The result is immediate from (11.3). $\square$

1. *Réflexions*, pp. 52–53.

Property 4. If $\mathcal{D}$ for a body of ideal gas is isothermally convex, the condition $g' = $ const. is equivalent to each and all of the conditions listed in Property 7.

Proof. From (11.4) we see that $g' = $ const. if and only if K_V is a function of θ alone. The result then follows from Property 7. □

Historical Comment. In the context of his own special theory CARNOT showed that K_V was a function of θ alone if and only if $g' = $ const. Property 4 asserts, among other things, that this conclusion of his is general, not restricted to the caloric theory of heat.

Remark 1. It is easy to see that Property 4 extends Property 4_{100}, stated at the end of Chapter 7, to the entire constitutive domain of the body of ideal gas. Indeed, from Property 4_{100} we may derive the following result for a body of ideal gas for which $\mathcal{D}$ (not necessarily a thermodynamic part) is isothermally convex and consists of thermodynamic points alone. Then γ *is a function of θ alone if and only if, in every Carnot web, $g'_{\phi_1\phi_2}(\theta)$ depends on ϕ_1 and ϕ_2 alone, not on θ, for all choices of the adiabats represented by ϕ_1 and ϕ_2.* It should be noted that this result, unlike Property 4, can be derived without use of CARNOT's General Axiom. However, when that axiom is imposed,

$$g'_{\phi_1\phi_2} = K_{\phi_1\phi_2} g', \tag{11.6}$$

$K_{\phi_1\phi_2}$ being a positive constant depending on ϕ_1 and ϕ_2. The result stated above, in conjunction with (11.6), leads to another proof of Property 4.

Remark 2. Since $\Lambda_V > 0$ throughout $\mathcal{D}$, all simple Carnot cycles have the shape and orientation indicated in Example (A) of Figure 9 in Chapter 7. Further, since $\mathcal{D}$ is a thermodynamic part, it is simply connected, and therefore every simple Carnot cycle is also ordinary. For ideal gases it is not necessary to distinguish between simple Carnot cycles and ordinary Carnot cycles.

Property 8. Let a simple Carnot cycle for a body of ideal gas be labelled as in Figure 11 of Chapter 7. Then

$$\frac{\log \dfrac{V_c}{V_d}}{\log \dfrac{V_b}{V_a}} = \frac{g'(\theta^-)}{g'(\theta^+)}. \tag{11.7}$$

The relation $V_b/V_a = V_c/V_d$ holds for every simple Carnot cycle if and only if $g' = $ const.

Proof. From (11.5) we see that for the simple Carnot cycle

$$C^+ = \frac{h(\theta^+)}{g'(\theta^+)} R \log \frac{V_b}{V_a}, \qquad C^- = \frac{h(\theta^-)}{g'(\theta^-)} R \log \frac{V_c}{V_d}. \tag{11.8}$$

By use of (10.6) in (11.8), (11.7) follows. If $g' = $ const., from (11.7) we see that $V_b/V_a = V_c/V_d$ for every simple Carnot cycle. Conversely, suppose the relation $V_b/V_a = V_c/V_d$ holds for every simple Carnot cycle. If B is a temperature corresponding to a point of $\mathcal{D}$, we can construct a Carnot web $\mathcal{W}_{A,C}$ such that $A < B < C$ (*cf.* Remark 1 after Definition 15 in Chapter 7). Applying (11.7) and the assumed relation to the cycles of the web, we conclude that g' is constant on $[A, C]$. Since B is an arbitrary temperature, g'' exists and is zero throughout the temperature interval of $\mathcal{D}$. Hence g' is constant, and the last statement of Property 8 follows. $\square$

Remark. We can also derive (11.7) from (7.15) and (11.6).

Property 3. Let $\mathcal{D}$ for a body of ideal gas be isothermally convex. Let a typical simple Carnot cycle be labelled as in Figure 11 of Chapter 7. Then γ is a function of θ alone if and only if the relation $V_b/V_a = V_c/V_d$ holds for every simple Carnot cycle.

Proof. The result follows by combining the last sentence of Property 8 with Property 4. $\square$

Remark. Property 3, like Property 4, is an extension of an earlier result, namely, Property 3_{loc} of Chapter 7, to the entire constitutive domain. By combining the result stated in Remark 1 after Property 4 with Properties 3_{loc} and 4_{loc}, without recourse to Axiom III, we arrive at the following result, similar to Property 3_{loc}: Let $\mathcal{D}$ (not necessarily a thermodynamic part) for a body of ideal gas be isothermally convex and consist of thermodynamic points alone. Then γ *is a function of θ alone if and only if $V_b/V_a = V_c/V_d$ for every ordinary Carnot cycle.* A direct proof of Property 3 may be based on this result and Remark 2 after Property 4.

Property 9. For a body of ideal gas MAYER's Assumption (3.14) is equivalent to

$$Jh = \theta g'. \tag{11.9}$$

Proof. The result is immediate from (11.3). $\square$

It might be conjectured that MAYER's Assumption (3.14) and the LAPLACE-POISSON Law (4.10) of Adiabatic Change would imply both specific heats of an ideal gas to be constant. We shall show now that this conjecture is true if $\mathcal{D}$ is the entire V-θ quadrant, though it is not true in general. We have already derived (4.14) as the general condition, as far as the Doctrine of Latent and Specific Heats is concerned, for the two assumptions to be compatible. We have just shown that the first of the assumptions is equivalent to (11.9). By substituting (11.2) into (4.14), we may show that *if $\mathcal{D}$ is isochorically convex, the functions Λ_V, K_V, and h given by (4.13) are necessary and sufficient,* within the theory based upon Axioms I–III, *for an ideal gas with a constant difference of specific heats to obey the* LAPLACE-POISSON *Law of Adiabatic Change.* Of course g is determined from h and J by (11.9). If $C \neq 0$, the specific heats are not constant.

If $\mathscr{D}$ is a region in which θ has a finite upper bound and V has a positive lower bound, we may give C any positive value we like in the second and third equations of (4.13) and then determine D and E so that $K_V > 0$ and $h > 0$. A similar condition on $\mathscr{D}$ yields the same inequalities if $C < 0$. On the contrary, if $\mathscr{D}$ is the entire quadrant, 0 is the only value of C compatible with the requirements $K_V > 0$, $h > 0$. Noting that $K_V = $ const. when $C = 0$, we conclude that *a body of ideal gas for which $\mathscr{D}$ is the entire V-θ quadrant is compatible with* MAYER'S *Assumption and the* LAPLACE-POISSON *Law of Adiabatic Change if and only if both the specific heats are constant.*

Historical Scholion. J. R. MAYER (1842) asserted positively his belief that heat and work were uniformly and universally interconvertible; he asserted also the relation (3.14) for an ideal gas and used it to calculate J, the mechanical equivalent of a unit of heat. On the basis of these two historical facts it seems to be widely believed that the latter assertion somehow reflects the former. It does not. Indeed, Property 9 shows MAYER'S Assumption (3.14) to be just as compatible with CARNOT'S theory as it is with CLAUSIUS'! For CARNOT'S theory we can choose $h = 1$, and (11.9) then determines the function g as follows:

$$g = J \log \theta + \text{const.} \tag{11.10}$$

This is no quibble. HOLTZMANN (1845), after asserting his belief in interconvertibility just as had MAYER before him, went on to contradict it tacitly in his calculations, which are all set within the caloric theory. Of course, since he assumed (3.13) also, which Property 2 in Chapter 3 shows to be equivalent to MAYER'S (3.14), his results are consistent with CARNOT'S theory if g has the form (11.10). Had MAYER been able to handle the mathematics of the caloric theory as HOLTZMANN was, he might well have gone on, as HOLTZMANN later did, to contradict in equations his assertion in words. MAYER'S innocence of mathematics, which kept him in the pure world of "physics", may be all that has earned him a place as a discoverer while denying even the rank of rediscoverer to HOLTZMANN.

The logical status of MAYER'S famous numerical value of J is made clear by comparing Property 9 with Corollary 10.3 of Chapter 10 and (11.2): *If* in all bodies of ideal gas (1) K_V is a function of temperature alone, (2) heat and work are uniformly interconvertible with the same mechanical equivalent of a unit of heat, and (3) MAYER'S Assumption (3.14) holds, *then* the positive constant J in MAYER'S Assumption is the common mechanical equivalent of a unit of heat in all bodies of ideal gas. Without the first two conditions in the foregoing statement, J may indeed be calculated in just the same way from the values of R, K_p, and K_V for a particular body of ideal gas at a particular volume and temperature, but it then need have no universal status even for all bodies of ideal gas.

According to Corollary 10.4 of Chapter 10, if J is a positive constant the two conditions $Jh = \theta g'$ and $g' = $ const. are together necessary and sufficient that a body of ideal gas shall operate in Carnot cycles with the classical efficiency according to (10.21). Using Properties 4 and 9 we read off:

Property 10. Let $\mathscr{D}$ for a body of ideal gas be isothermally convex, and let J be a positive constant. Then the conditions $K_V = k(\theta)$ and $J(K_p - K_V) = R$ are necessary and sufficient that

$$g = Jh + \text{const.}, \qquad h = M\theta, \qquad M = \text{const.} > 0, \qquad (10.22)_r$$

and hence that the gas achieve the classical efficiency $1 - \theta^-/\theta^+$ in every Carnot cycle whose operating temperatures are θ^+ and θ^-; equivalently, necessary and sufficient that the function G in (8.1) be of the form

$$G(x, y, z) = J(1 - y/x)z. \qquad (11.11)$$

In particular, if the gas has constant specific heats and if J is such as to satisfy (3.14), then (10.22) and (11.11) hold, and the gas attains the classical efficiency in every Carnot cycle. It is easy to verify that the previous statement remains true if we set aside the restriction that $\mathscr{D}$ be isothermally convex.

General Scholion. In all the foregoing analysis the functions g and h and also the constant J in MAYER's Assumption (3.14) are regarded as being possibly constitutive. According to Property 10, if a particular body of ideal gas obeys MAYER's Assumption and has a specific heat that is a function of temperature alone, then the functions g and h for it have the form (10.22). Clearly, if ideal gases are to attain the classical efficiency in all Carnot cycles, their functions g and h *must* be of the form (10.22). It does not follow that J has the same value for all such bodies of ideal gases. Furthermore, no claim has yet been made that g and h must have the form (10.22) for other bodies. Indeed, no such claim can be made except on the basis of further axioms. The stage is now set for those axioms, but we shall defer their statement until Chapter 15 so as first to obtain further consequences of Axioms I–III. In this way we shall complete REECH's program of constructing a thermodynamics that includes as special cases both CARNOT's theory and CLAUSIUS', so the two may be compared and contrasted at each stage.

Property 11. Let one, and hence both, of the specific heats of a body of ideal gas whose constitutive domain is isothermally or isochorically convex be functions of temperature alone. Then there are a positive constant c and a constant d such that $g = c\theta + d$, and

$$E_{c\theta + d, h} = \frac{Rd}{c} \log V + \int \frac{(c\theta + d)K_V(\theta)}{h(\theta)} d\theta. \qquad (11.12)$$

In particular, there is a choice of g, namely $g = c\theta$, such as to render the corresponding internal pro-energy $E_{c\theta, h}$ a function of temperature alone. Conversely, if for a body of ideal gas there is a choice of the functions g and h such as to render $E_{g, h}$ a function of temperature alone, then the specific heats are also functions of temperature alone for the body, and $g = c\theta$, c being a positive constant.

Proof. To obtain these results, we use (9.27), (11.1), (11.2), and the fact that g is an increasing function. □

Property 12. Let $\mathscr{D}$ for a body of ideal gas be isothermally convex, and let J be a positive constant. Then MAYER's Assumption (3.14) and the requirement that $E_{g,h}$ be a function of θ alone are necessary and sufficient conditions that

$$g = Jh, \qquad h = M\theta, \tag{11.13}$$

M being a positive constant. The conditions (11.13) are sufficient but not necessary to ensure that every Carnot cycle achieve the classical efficiency.

Proof. The results are immediate from Property 9, Property 11, and (11.2). The last statement follows from Corollary 10.4 of Chapter 10. □

Historical Scholion. CLAUSIUS in his first paper on thermodynamics, which referred mainly to ideal gases, in effect laid down three axioms:

1. The uniform interconvertibility of heat and work in cyclic processes.
2. HOLTZMANN's Assumption (3.13), J being the mechanical equivalent of a unit of heat.
3. CARNOT's General Axiom (8.1), G being a universal function, with the inessential difference that $C^-(\mathscr{C})$ replaces $C^+(\mathscr{C})$.

On this basis he proved the existence of an internal energy of a body of ideal gas and showed that it could be taken as a function of θ alone. From his formulae he could easily have inferred at once the expression for classical efficiency in Carnot cycles, as KELVIN showed a little later. We can recover CLAUSIUS' results trivially from the structure given here. Indeed, by Corollary 10.3 in Chapter 10 his first assumption amounts to

$$g = Jh + \text{const.} \tag{10.16_r}$$

By use of Property 2 in Chapter 3 and Property 9 we may write his second assumption as

$$Jh = \theta g'. \tag{11.9_r}$$

Hence (10.22) follows, and with it (10.21) for Carnot cycles and all the other formulae of classical thermodynamics. CLAUSIUS' assertion about the internal energy amounts to Property 11. CLAUSIUS' only application of his third assumption was similar to CARNOT's own use of his General Axiom so as to derive (9.22). However, since CLAUSIUS' second assumption is itself the special case of (9.22) corresponding to the choice $C(\theta) = J/\theta$, his results for ideal gases do not require his third assumption at all. The analysis we have given above has followed a different route and delivers, among many other results, a converse to CLAUSIUS', by use of Corollary 10.4: If CARNOT's

General Axiom is accepted, then CLAUSIUS' first two assumptions for ideal gases are necessary as well as sufficient in order that the efficiency of bodies of ideal gas in Carnot cycles shall be $1 - \theta^-/\theta^+$. In the Appendix to Chapter 15 we submit the ideas expressed by CLAUSIUS to further scrutiny.

Remark. Our development has separated the Doctrine of Latent and Specific Heats from the far more specific theory called thermodynamics. The former, set forth by Axiom II, sufficed in conjunction with Axiom I to obtain Properties 1, 2, 3_{loc}, and 4_{loc}. Thermodynamics, making use of Axiom III as well as Axioms I and II, delivered Properties 3–12. Now disregarding this distinction, we may summarize the main properties of ideal gases as follows. From Properties 3, 4, 7, and 11 we see that *for a body of ideal gas whose constitutive domain is isothermally convex the six following conditions are equivalent:*

1. K_V is a function of θ alone.
2. K_p is a function of θ alone.
3. γ is a function of θ alone.
4. $g' = $ const.
5. $V_b/V_a = V_c/V_d$ for every simple Carnot cycle.
6. There is a choice of g and h such as to make $E_{g,h}$ a function of θ alone.

Properties 2 and 9 show that *the following three conditions are equivalent*:

7. MAYER's Assumption (3.14).
8. HOLTZMANN's Assumption (3.13).
9. $Jh = \theta g'$.

By replacing the conditions in Property 10 by equivalent ones from the two lists above, we can obtain eighteen pairs of hypotheses necessary and sufficient to entail the classical efficiency in Carnot cycles.

Van der Waals Fluids

Application of (9.2), (9.19), and (9.20) to a body of Van der Waals fluid, defined by the thermal equation of state (2.10), shows that for such a fluid g' is a continuously differentiable positive function, that

$$\Lambda_V = \frac{Rh}{(V - b)g'} > 0, \qquad \frac{\partial K_V}{\partial V} = -\frac{Rhg''}{(V - b)g'^2}, \qquad (11.14)$$

and, in a part of the constitutive domain that is isothermally convex, that

$$K_V = -\frac{Rhg''}{g'^2} \log(V - b) + K, \qquad (11.15)$$

K being a function of θ alone. From the second equation of (11.14) we see that *a necessary condition for K_V to be a function of θ alone is $g' = $ const.* This condition is also sufficient in an isothermally convex part of $\mathscr{D}$.

We have discussed the isothermal convexity of $\mathscr{D}$ in Footnote 1 to Chapter

9. Let us now suppose that $g' = $ const. and then apply the conclusion reached there. We find that K_V may be either of two continuously differentiable, positive functions of θ, say K_V^1 and K_V^2, satisfying the requirement that they be one and the same function when $8a/(27Rb) \leq \theta < \infty$ and have the same derivatives when $\theta = 8a/(27Rb)$. For this body, along an adiabat

$$(V - b)^{R/J} \exp \int \frac{K_V(\theta)}{h(\theta)} d\theta = \text{const.}, \tag{11.16}$$

J being a constant equal to g'. One choice of K_V gives the adiabats in the region where $V < 3b$; the other, where $V > 3b$.

Of course we do not have to choose K_V^1 and K_V^2 as being distinct functions. If we ask that both be constant, then the requirement of continuity forces both to be the same constant. If also $h = \theta$, (11.16) reduces to (5.3). The formal similarity between (5.3) and the familiar LAPLACE–POISSON Law (4.10) is deceptive, for γ is not a constant unless $a = 0$, and only in that case (in fact excluded by definition) is $\gamma - 1 = R/(JK_V)$.

Consider again a Van der Waals fluid for which $g' = $ const. For such a fluid there are a positive constant c and a constant d such that $g = c\theta + d$, and by use of (11.14) and (9.27) we show that when $\theta \geq 8a/(27Rb)$

$$E_{c\theta+d,h} = \frac{Rd}{c} \log(V - b) - \frac{a}{V} + \int_{8a/(27Rb)}^{\theta} \frac{(cu + d)K_V(u)}{h(u)} du + \text{const.} \tag{11.17}$$

Also the same expression delivers the two branches of $E_{c\theta+d,h}$ when $0 < \theta < 8a/(27Rb)$, one for each choice of K_V, the constant in (11.17) being the same for both branches. Since $a > 0$, there is no choice of g and h such as to render $E_{g,h}$ a function of θ alone even if $\theta \geq 8a/(27Rb)$ (cf. Property 11 of an ideal gas).

CHAPTER 12

Relation of Motors to Refrigerators.

Scholion. Hitherto we have emphasized cycles which can be regarded as models for heat engines, namely, cycles that absorb heat so as to give out work. The *motive efficiency* $e_{\text{mot}}(\mathscr{C})$ of a cycle $\mathscr{C}$ which absorbs heat is defined by

$$e_{\text{mot}}(\mathscr{C}) \equiv \frac{L(\mathscr{C})}{JC^+(\mathscr{C})}, \tag{12.1}$$

J denoting a positive constant such as to render the right-hand side of (12.1) dimensionless. The motive efficiency was called simply "efficiency" in previous chapters. Although for a Carnot cycle in a thermodynamic part $\mathscr{D}_{\text{th}}$ the motive efficiency is always positive, as shown by (10.7), for other cycles it may be zero or negative or fail to exist. For example, the reverse of a Carnot cycle in $\mathscr{D}_{\text{th}}$ consumes work and gives out heat, thus serving as a cooler rather than a motive engine. Indeed, for a C-cycle in $\mathscr{D}_{\text{th}}$ it follows from (10.7) that the motive efficiency $\gtreqless 0$ if and only if $\theta^+ \gtreqless \theta^-$, respectively, θ^+ and θ^- being the absorption temperature and the emission temperature.

The *refrigerant efficiency* or "coefficient of performance" of a cycle $\mathscr{C}$ such that $L(\mathscr{C}) \neq 0$ is defined by

$$e_{\text{ref}}(\mathscr{C}) \equiv \frac{JC^+(\mathscr{C})}{-L(\mathscr{C})}. \tag{12.2}$$

Lemma. *Let $\mathscr{C}$ be any cycle such that $C^+(\mathscr{C}) > 0$ and $L(\mathscr{C}) \neq 0$. Then*

$$e_{\text{mot}}(\mathscr{C})e_{\text{ref}}(-\mathscr{C}) = \frac{C^-(\mathscr{C})}{C^+(\mathscr{C})} = \frac{C^+(-\mathscr{C})}{C^-(-\mathscr{C})} = \frac{C^-(\mathscr{C})}{C^-(-\mathscr{C})} = \frac{C^+(-\mathscr{C})}{C^+(\mathscr{C})}, \tag{12.3}$$

and

$$e_{\text{ref}}(-\mathscr{C}) - \left[\frac{1}{e_{\text{mot}}(\mathscr{C})} - 1\right] = 1 - \frac{1}{e_{\text{mot}}(\mathscr{C})}\left[1 - \frac{C^{-}(\mathscr{C})}{C^{+}(\mathscr{C})}\right]. \qquad (12.4)$$

Proof. The Reversal Theorems expressed by (2.14) and (3.5) yield (12.3) immediately. To obtain (12.4), we begin with the left-hand side of it and then eliminate $e_{\text{ref}}(-\mathscr{C})$ by use of the first equation of (12.3). □

Theorem 12. *Let $\mathscr{C}$ denote any cycle such that $C^{+}(\mathscr{C}) > 0$ and $L(\mathscr{C}) \neq 0$. Then the condition*

$$e_{\text{ref}}(-\mathscr{C}) = \frac{1}{e_{\text{mot}}(\mathscr{C})} \qquad (12.5)$$

for every cycle $\mathscr{C}$ is necessary and sufficient that $C^{+}(\mathscr{C}) = C^{-}(\mathscr{C})$ for every cycle $\mathscr{C}$. In $\mathscr{D}_{\text{th}}$ the condition (12.5) or, equivalently, its restriction to ordinary Carnot cycles, is also equivalent to any of the three conditions listed in Corollary 9.3 of Chapter 10. If J is the constant in (12.1), then the condition

$$e_{\text{ref}}(-\mathscr{C}) = \frac{1}{e_{\text{mot}}(\mathscr{C})} - 1 \qquad (12.6)$$

for every cycle $\mathscr{C}$ is necessary and sufficient that $L(\mathscr{C}) = JC(\mathscr{C})$ for every cycle $\mathscr{C}$. In $\mathscr{D}_{\text{th}}$ the condition (12.6) or its special case when $\mathscr{C}$ is restricted to ordinary Carnot cycles alone is also equivalent to any of the three conditions listed in Corollary 10.3 of Chapter 10.

Proof. These results follow from (12.1), (12.4), and Corollaries 9.3 and 10.3 of Chapter 10. □

Remark 1. This theorem enables us to characterize the two principal theories of heat essentially in terms of the refrigerators obtained by running motors backward. According to CARNOT's theory a motor of efficiency near to 1 becomes, if run backward, a refrigerator of efficiency near to 1, as is shown by (12.5). According to CLAUSIUS' theory, on the contrary, a nearly perfect motor if run backward becomes an almost ineffectual refrigerator, as is shown by (12.6).

Remark 2. If $\mathscr{C}$ is a Carnot cycle in $\mathscr{D}_{\text{th}}$, by using (10.6) in the first equation of (12.3) we see that

$$e_{\text{mot}}(\mathscr{C})e_{\text{ref}}(-\mathscr{C}) = \frac{h(\theta^{-})}{h(\theta^{+})}, \qquad (12.7)$$

θ^{+} and θ^{-} being the operating temperatures of $\mathscr{C}$. This formula provides a further general interpretation for the heat-loss function h.

EXAMPLE. Consider a body of ideal gas whose constitutive domain is a thermodynamic part. If the gas obeys MAYER's Assumption (3.14) and has specific heats that are functions of θ alone, Property 10 in Chapter 11 shows that $0 < e_{\text{mot}}(\mathscr{C}) = 1 - \theta^-/\theta^+ < 1$ in every Carnot cycle $\mathscr{C}$. By (12.7), if $\mathscr{C}$ is a Carnot cycle whose operating temperatures are θ^+ and θ^-,

$$0 < e_{\text{ref}}(-\mathscr{C}) = \frac{1}{(\theta^+/\theta^-) - 1} < \infty, \tag{12.8}$$

since (10.22) holds for the body of ideal gas.

CHAPTER 13

Estimates of the Efficiency of a Body undergoing a Cyclic Process.

In the preceding chapter motive and refrigerant efficiencies were defined only for cycles. However, the same definitions can be used for bodies undergoing processes that need not be cyclic. Henceforth we shall refer to the motive efficiency of a body undergoing a process simply as the efficiency of the process.

Lemma. *Let a be an arbitrary constant. For a cyclic process in $\mathcal{D}_{th}$ or for any simple cyclic process in $\mathcal{D}_n$,*

$$L = aC^+ - \int_{\mathcal{J}^+} \left(a - \frac{g}{h}\right)Q dt - \int_{\mathcal{J}^-} \frac{g}{h}(-Q)dt. \qquad (13.1)$$

Proof. The result is immediate from (10.8). That formula is justified for cyclic processes in $\mathcal{D}_{th}$ by Lemma 2 of Chapter 10; for simple cycles in $\mathcal{D}_n$, by appeal to Lemma 1. $\square$

Remark 1. Clearly we may replace g and h in (13.1) by any $\bar{g}$ and $\bar{h}$ related to them through (9.17).

Remark 2. The identity (13.1), like (6.11), is similar to one obtained by Messrs. FOSDICK & SERRIN[1] in their research on continuum thermodynamics.

1. R. L. FOSDICK & J. SERRIN, "Global properties of continuum thermodynamic processes", *Archive for Rational Mechanics and Analysis* **59** (1975), 97–109. See also C. TRUESDELL, "Improved estimates of the efficiencies of irreversible heat engines", *Annali di Matematica Pura ed Applicata* (IV) **108** (1976), 305–323, and C. TRUESDELL, "Irreversible heat engines and the second law of thermodynamics", *Letters in Heat and Mass Transfer* **3** (1976), 267–290.

With the help of (13.1) it is easy to derive estimates for L by making judicious choices of the constant a and the function g, just as the emission–absorption estimates were derived in Chapter 6 from (6.11).

Theorem 13 (Efficiency Estimates). *Let a cyclic process in $\mathscr{D}_{\mathrm{th}}$ or a simple cyclic process in $\mathscr{D}_{\mathrm{n}}$ have $\theta_{\max}$ and $\theta_{\min}$ as its maximum and minimum temperatures. Then*

$$\left[\min \frac{g - g(\theta_{\max})}{h}\right] C^+ \leq L \leq \left[\max \frac{g - g(\theta_{\min})}{h}\right] C^+. \tag{13.2}$$

The upper bound is achieved if and only if

$$\theta = \theta_{\min} \quad \text{on } \mathscr{T}^-, \qquad \frac{g(\theta) - g(\theta_{\min})}{h(\theta)} = \max \frac{g - g(\theta_{\min})}{h} \quad \text{on } \mathscr{T}^+.$$

The lower bound is achieved if and only if

$$\theta = \theta_{\max} \quad \text{on } \mathscr{T}^-, \qquad \frac{g(\theta) - g(\theta_{\max})}{h(\theta)} = \min \frac{g - g(\theta_{\max})}{h} \quad \text{on } \mathscr{T}^+.$$

If $\mathscr{T}^+$ and $\mathscr{T}^-$ are not empty,

$$\inf_{\mathscr{T}^+}\left[\frac{g - g\left(\sup_{\mathscr{T}^-} \theta\right)}{h}\right] \leq \frac{L}{C^+} \leq \sup_{\mathscr{T}^+}\left[\frac{g - g\left(\inf_{\mathscr{T}^-} \theta\right)}{h}\right]. \tag{13.3}$$

The upper bound is achieved if and only if the lower one is; equivalently, if and only if θ has a constant value, say θ^-, on $\mathscr{T}^-$, and $[g - g(\theta^-)]/h$ has a constant value on $\mathscr{T}^+$.

Proof. First we replace g by $g - g(\theta_{\min})$ and set

$$a = \max[g - g(\theta_{\min})]/h$$

in (13.1). Since the integrands in (13.1) are then nonnegative, the upper bound in (13.2) follows. Further, the upper bound is achieved if and only if both of the integrals in (13.1) vanish; that is, if and only if on $\mathscr{T}^-$, $\theta = \theta_{\min}$, and on $\mathscr{T}^+$, $[g - g(\theta_{\min})]/h = \max[g - g(\theta_{\min})]/h$. Replacing g by $g - g(\theta_{\max})$ and setting $a = \min[g - g(\theta_{\max})]/h$ in (13.1), we obtain the lower bound in (13.2) and read off necessary and sufficient conditions that that bound be attained. The rest of the theorem follows by similar reasoning. $\square$

Remark 1. The bounds in (13.2) and (13.3) are independent of the choice of g and h allowed by the theory; that is, if in them g and h are replaced by any $\bar{g}$ and $\bar{h}$ satisfying (9.17), the values of the bounds remain unchanged. It is also easy to verify that the upper bound in (13.2) is always nonnegative while the lower bound is always nonpositive. If g and h are known and if $C^+ > 0$, (13.2) enables us to calculate upper and lower bounds for L/C^+ from a knowledge of the maximum and minimum temperatures alone.

Remark 2. To arrive at the estimate (13.3), it has been assumed that $\mathscr{T}^+$ and $\mathscr{T}^-$ are not empty for the cyclic process. However, this restriction is trivial. From Lemma 3 of Chapter 10 or from the remark following it we see that for a cyclic process in $\mathscr{D}_{th}$ either both or neither of the sets $\mathscr{T}^+$ and $\mathscr{T}^-$ is empty. In the former case, the process is adiabatic, the work done is zero, and both bounds in (13.2) are attained trivially.

Remark 3. The efficiency theorem for C-processes (Theorem 10 in Chapter 10) may be proved as an immediate consequence of the necessary and sufficient conditions for equality to hold in (13.3).

Remark 4. Though the estimate (13.3) is usually sharper than the estimate (13.2), it is less useful, since it requires knowledge about the sets $\mathscr{T}^+$ and $\mathscr{T}^-$. While necessary and sufficient conditions that the bounds be attained have been stated in Theorem 13, the following simpler sufficient conditions for C-processes may be of some interest. A C-process in $\mathscr{D}_{th}$ achieves both bounds in (13.3). A C-process which emits heat at its lowest temperature and absorbs heat at a temperature which maximizes $[g - g(\theta_{min})]/h$ on $[\theta_{min}, \theta_{max}]$ achieves the upper bound in (13.2). A C-process which emits heat at its highest temperature and absorbs heat at a temperature which minimizes $[g - g(\theta_{max})]/h$ on $[\theta_{min}, \theta_{max}]$ achieves the lower bound in (13.2). To see that such processes need not be the only ones to achieve the bounds, we need only note that the functions $[g - g(\theta_{min})]/h$ and $[g - g(\theta_{max})]/h$ may attain their maximum or minimum values throughout some subinterval of $[\theta_{min}, \theta_{max}]$.

Remark 5 (Application to the Principal Theories of Heat). In the caloric theory we may take h as 1 and so reduce the estimate (13.2) to

$$[g(\theta_{min}) - g(\theta_{max})]C^+ \leqq L \leqq [g(\theta_{max}) - g(\theta_{min})]C^+; \qquad (13.4)$$

likewise, (13.3) reduces to

$$g\left(\inf_{\mathscr{T}^+} \theta\right) - g\left(\sup_{\mathscr{T}^-} \theta\right) \leqq \frac{L}{C^+} \leqq g\left(\sup_{\mathscr{T}^+} \theta\right) - g\left(\inf_{\mathscr{T}^-} \theta\right). \qquad (13.5)$$

If, on the other hand, the body interconverts heat and work uniformly, (10.16) holds, and since h is then an increasing function, (13.2) reduces to

$$J\left[1 - \frac{h(\theta_{max})}{h(\theta_{min})}\right]C^+ \leqq L \leqq J\left[1 - \frac{h(\theta_{min})}{h(\theta_{max})}\right]C^+, \qquad (13.6)$$

while (13.3) reduces to

$$J\left[1 - \frac{h\left(\sup_{\mathscr{T}^-} \theta\right)}{h\left(\inf_{\mathscr{T}^+} \theta\right)}\right] \leqq \frac{L}{C^+} \leqq J\left[1 - \frac{h\left(\inf_{\mathscr{T}^-} \theta\right)}{h\left(\sup_{\mathscr{T}^+} \theta\right)}\right]. \qquad (13.7)$$

Confining attention to processes which absorb heat, we note that the upper bound in (13.4) or (13.6) is achieved only by C-processes which absorb heat at their maximum temperature and emit it at their minimum temperature, while the lower bound in (13.4) or (13.6) is achieved only by C-processes which absorb heat at their minimum temperature and emit it at their maximum temperature. In (13.5) and (13.7) the upper bound or lower bound is achieved by C-processes and by them alone.

Historical Comment. From the second inequality of (13.2) and the second inequality of (10.3) it follows that if $L > 0$ for a cyclic process in $\mathscr{D}_{\mathrm{th}}$, then $C^+ > 0$, $C^- > 0$, and $\theta_{\max} > \theta_{\min}$. Thus a cyclic process delivering motive power cannot be isothermal, nor can it fail to emit heat as well as absorb heat. CARNOT himself wrote (*Réflexions*, p. 11): "... to give rise to motive power it is not enough to produce heat; cold also must be produced; without it, the heat would be useless". This striking statement has sometimes been interpreted as foreshadowing the "Second Law of Thermodynamics". We have seen that this claim of CARNOT is valid not only in the caloric theory but also in the more general theory based on CARNOT's General Axiom.

Both from the second inequality of (13.5) and from the second inequality of (13.7) we see also that if $L > 0$, then $\sup_{\mathscr{I}^+} \theta > \inf_{\mathscr{I}^-} \theta$. That is, according to the two principal theories of heat, in order for a body undergoing a cyclic process to do positive work, some of the temperatures at which it absorbs heat must be higher than some at which it emits heat. Although something of the sort was claimed by CARNOT (*Réflexions*, p. 9), the explicit statement is one of the "Second Laws" attributed commonly to CLAUSIUS.

Corollary 13.1 (Maximum and Minimum Efficiency). *Consider the collection of all cyclic processes in $\mathscr{D}_{\mathrm{th}}$ that absorb heat and have a common maximum temperature $\theta_{\max}$ and a common, lower, minimum temperature $\theta_{\min}$. Let $[g - g(\theta_{\min})]/h$ be an increasing function on the interval $[\theta_{\min}, \theta_{\max}]$. Then the efficiency of any process of this collection never exceeds the efficiency of a Carnot process in $\mathscr{D}_{\mathrm{th}}$ whose operating temperatures are $\theta_{\max}$ and $\theta_{\min}$; if the collection includes a Carnot process having $\theta_{\max}$ and $\theta_{\min}$ as its operating temperatures, the maximum efficiency for processes in the collection is attained by such Carnot processes and by them alone. If, on the other hand, $[g - g(\theta_{\max})]/h$ is an increasing function, the minimum efficiency consonant with the given extremes of temperature is that of the reverse of a Carnot process, if such exists with those temperatures as its operating temperatures. Corresponding statements hold for the collection of simple cyclic processes in $\mathscr{D}_{\mathrm{n}}$.*

> *Proof.* Recalling (10.7), we see that the assertion about maximum efficiency follows from the upper bound in (13.2) and the condition that that bound be attained, as stated in Theorem 13. The assertion about minimum efficiency follows from parallel consideration of the lower bound in (13.2). □

Remark. The property of Carnot processes set forth carefully in the statement of Corollary 13.1 is that which is commonly and loosely declared as follows: *Carnot cycles, and they alone, achieve maximum efficiency for given extremes of temperature; the reverses of Carnot cycles, and they alone, achieve minimum efficiency.* We shall use these terms to state the next corollary.

Corollary 13.2. *Carnot cycles, and those cycles alone, achieve maximum efficiency for bodies of two kinds:*

1. *Bodies that obey the caloric theory.*
2. *Bodies that interconvert heat and work uniformly.*

For these bodies the reverses of Carnot cycles, and they alone, achieve minimum efficiency.

> *Proof.* For bodies of either kind, both $[g - g(\theta_{\min})]/h$ and $[g - g(\theta_{\max})]/h$ are increasing functions. $\square$

General Scholion. Assertion 1 in the corollary just stated is traditionally attributed to CARNOT, and something of the sort may be inferred from the tantalizing obscurities of his treatise. So far as we know, the first argument to this effect that could be regarded as a proof is that given in TRUESDELL's paper cited above in Footnote 5 to Chapter 2. That proof is based directly on some formulae which KELVIN derived from the caloric theory and which have no useful counterparts in CLAUSIUS' theory. The proof in the text above, descending from an inclusive estimate valid for theories satisfying CARNOT's General Axiom and itself based upon a method introduced by Messrs. FOSDICK & SERRIN, is new. Perhaps it will stand as the last addition to the caloric theory of heat.

Although Assertion 2 of Corollary 13.2, specialized to classical thermodynamics, may be found in every textbook and is supported therein by an argument deriving from CARNOT through works of CLAPEYRON, KELVIN, RANKINE, and CLAUSIUS, that argument is not a proof. The first proof, so far as we know, is that given incidentally in TRUESDELL's paper, "The efficiency of a homogeneous heat engine", *Journal of Mathematical and Physical Sciences* (Madras) **7** (1973), 349–371, **9** (1975), 193–194; the results in that paper were announced in "Sul rendimento delle macchine termiche omogenee", *Rendiconti dell'Accademia Nazionale dei Lincei, Classe di Scienze Fisiche Matematiche e Naturali* (8) **53** (1973), 549–553. The theory given there concerns possibly irreversible processes in which energy is balanced.

The classical inference that Carnot cycles are the most efficient, we have given in CARNOT's own words in Footnote 6 to the Preface of this tractate. It makes no use of any particular theory of thermodynamics and was taken over with no real changes by CLAUSIUS, KELVIN, and later writers. Some version of it may be found in every textbook today. Corollary 13.2 shows that the result asserted by CARNOT and CLAUSIUS is true in the theories proposed by them and also in all theories in which heat and work are supposed to be uniformly interconvertible, provided Axioms I–III be accepted. Thus it might seem that the classical claim were true even in the theory based on CARNOT's General Axiom. Such is not the case. We note that to arrive at the classical result in Corollary 13.1, the assumption that $\theta_{\max}$ be the only temperature in $[\theta_{\min}, \theta_{\max}]$ at which $[g - g(\theta_{\min})]/h$ attains its maximum is crucial. We shall show now that the classical claim is generally false by exhibiting an example of a fluid body which is compatible with Axioms I–III and for which $[g - g(\theta_{min})]/h$ does not attain its maximum at θ_{max}.

For all processes in the collection considered in Corollary 13.1, it follows from the second inequality in (13.2) that

$$\frac{L}{C^+} \leqq \max_{[\theta_{min}, \theta_{max}]} \left(\frac{g - g(\theta_{min})}{h} \right), \tag{13.8}$$

while for a Carnot process with operating temperatures θ_{max} and θ_{min}

$$\frac{L}{C^+} = \frac{g(\theta_{max}) - g(\theta_{min})}{h(\theta_{max})}. \tag{13.9}$$

Consider the example of a fluid body compatible with Axioms I–III, which was introduced at the beginning of Remark 2 after Corollary 11.2 in Chapter 10. For that example, choosing $\theta_{min} = 1$ and $\theta_{max} = 3$, we find that $[g(\theta_{max}) - g(\theta_{min})]/h(\theta_{max}) = \frac{2}{9}$ and $\max_{[1,3]}[g - g(\theta_{min})]/h = \frac{1}{4}$, the maximum being attained at $\theta = 2$ and for no other θ. Thus a Carnot process with operating temperatures θ_{max} and θ_{min} *does not achieve the upper bound in* (13.8). On the other hand, application of Theorem 13 shows that for this example the upper bound in (13.8) is achieved only by Carnot processes with operating temperatures 2 and 1. Nonsimple Carnot processes of this kind actually exist for the body in this example, as is indicated by the full lines in Figure 15. Indeed Carnot processes operating between the maximum and minimum temperatures 3 and 1 also exist. This example shows that in the theory based on Axioms I–III, Carnot processes with operating temperatures θ_{max} and θ_{min} are not always the most efficient processes. By slightly modifying the process in Figure 15 as indicated by the dashed lines we may construct cyclic processes that are not Carnot processes yet achieve efficiencies

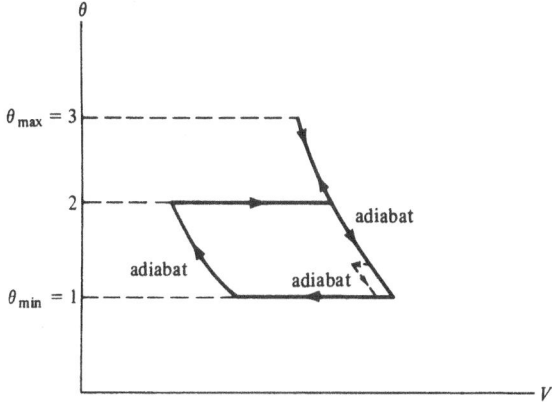

Figure 15. Example to illustrate the fact that CARNOT's General Axiom does not always make Carnot cycles achieve maximum efficiency.

arbitrarily close to the maximum efficiency, and that exceed in efficiency Carnot processes operating between $\theta_{\max}$ and $\theta_{\min}$. *Therefore the classical assertion in Corollary 13.2 cannot be extended to all theories compatible with* CARNOT'S *General Axiom, and the classical argument, which seems not to mention or use any special theory of heat, must be actually wrong, not merely unconvincing*—wrong, that is, if the classical words do in fact express an argument rather than merely a belief.

CHAPTER 14

The Fourth Theorem of Reech: Existence of Four Thermodynamic Potentials.

In the theory developed in this tractate a fluid is characterized by three *constitutive functions*: ϖ, Λ_V, and K_V, all three of these being functions of V and θ on $\mathscr{D}$. Here we consider only a thermodynamic part $\mathscr{D}_{th}$, defined by Definition 17 in Chapter 9.

Definition 19. A *thermodynamic potential* for a body on $\mathscr{D}_{th}$ is a single function that determines the restrictions to $\mathscr{D}_{th}$ of the constitutive functions ϖ, Λ_V, and K_V, on the presumption that g and h are known.

Remark. If g and h are known, and if the constitutive domain is a thermodynamic part, a thermodynamic potential is a single function that summarizes all the constitutive properties of a body that thermodynamics represents. The pressure function ϖ is not a thermodynamic potential: Although it determines Λ_V through (9.2), (9.1) shows that it does not determine K_V.

A body has infinitely many thermodynamic potentials on $\mathscr{D}_{th}$. We proceed to define and describe four of them. These four, although generally attributed to MASSIEU and others, were discovered by REECH.

Definition 20 (REECH, in essence). On a thermodynamic part $\mathscr{D}_{th}$ let $\Phi_{g,h}$, $X_{g,h}$, and $Z_{g,h}$ be defined as follows:

$$\begin{aligned}
\Phi_{g,h} &\equiv E_{g,h} - gH_h, \\
X_{g,h} &\equiv E_{g,h} + \varpi V, \\
Z_{g,h} &\equiv X_{g,h} - gH_h.
\end{aligned} \qquad (14.1)$$

The functions $\Phi_{g,h}$, $X_{g,h}$, and $Z_{g,h}$ are named the *free energy*, *enthalpy*, and *free enthalpy*, respectively, of the body on $\mathcal{D}_{\text{th}}$.

Remark 1.

$$E_{g,h} + Z_{g,h} = X_{g,h} + \Phi_{g,h}. \tag{14.2}$$

Remark 2. The notations indicate dependence on the choice of the functions g and h. If another choice $\bar{g}$ and $\bar{h}$ satisfying (9.17) is made, it is easy to work out the relations between $\Phi_{\bar{g},\bar{h}}$, $X_{\bar{g},\bar{h}}$, $Z_{\bar{g},\bar{h}}$ and $\Phi_{g,h}$, $X_{g,h}$, $Z_{g,h}$ by using (9.17) and (9.31). In the rest of this chapter, on the understanding that a fixed choice of g and h has been made once and for all, we shall omit the subscripts.

Lemma (REECH). *For a process in $\mathcal{D}_{\text{th}}$, almost always*

$$\begin{aligned}
\dot{E} &= g\dot{H} - \varpi\dot{V}, \\
-\dot{\Phi} &= \varpi\dot{V} + H\theta g', \\
\dot{X} &= g\dot{H} + V\dot{\varpi}, \\
\dot{Z} &= V\dot{\varpi} - H\theta g'.
\end{aligned} \tag{14.3}$$

Proof. These results follow from (9.28) and the definitions (14.1). □

Using the notation (9.32), by integrating the second equation of (14.3) from t_1 to t_2 we obtain the following

Corollary. *In an isothermal process in $\mathcal{D}_{\text{th}}$*

$$L = -\Delta\Phi. \tag{14.4}$$

That is, in an isothermal process a body does work at the expense of its free energy.

Notation For This Chapter Only

An overbar upon a letter will denote a function of V and θ. The functions introduced in the earlier parts of this tractate will not be exempt from this convention. For example, the functions denoted up to now by ϖ, Λ_V, K_V, g, h, and H_h will be denoted here by $\bar{\varpi}$, $\bar{\Lambda}_V$, $\bar{K}_V$, $\bar{g}$, $\bar{h}$, and $\bar{H}$, respectively. Thus, for example,

$$\Phi = \bar{\Phi}(V, \theta) = E_{g,h}(V, \theta) - g(\theta)H_h(V, \theta) = \bar{E}(V, \theta) - \bar{g}(\theta)\bar{H}(V, \theta). \tag{14.5}$$

We are now ready to state four propositions which we shall label Theorems 14A, 14B, 14C, and 14D. These four together express REECH's *Fourth Theorem*.

Theorem 14A.   *On $\mathscr{D}_{\text{th}}$*

$$\bar{\varpi} = -\frac{\partial \bar{\Phi}}{\partial V}, \qquad \bar{g}' \bar{H} = -\frac{\partial \bar{\Phi}}{\partial \theta}. \tag{14.6a}$$

At all points of $\mathscr{D}_{\text{th}}$ where $\bar{g}' > 0$,

$$\bar{\Lambda}_V = -\frac{\bar{h}}{\bar{g}'} \frac{\partial^2 \bar{\Phi}}{\partial V \partial \theta}, \qquad \bar{K}_V = -\bar{h} \frac{\partial}{\partial \theta} \left(\frac{1}{\bar{g}'} \frac{\partial \bar{\Phi}}{\partial \theta} \right). \tag{14.6b}$$

Proof.   These relations follow immediately from (9.26), (9.27), and the first equation of (14.1).   □

Remark.   Suppose $\bar{g}$ and $\bar{h}$ be known. Then from (14.6) it follows that the function $\bar{\Phi}$ determines $\bar{\varpi}$ in $\mathscr{D}_{\text{th}}$ and determines $\bar{\Lambda}_V$ and $\bar{K}_V$ at all points of $\mathscr{D}_{\text{th}}$ where $\bar{g}' > 0$. However, since $\bar{g}' > 0$ except on a set of temperatures with empty interior, every point of $\mathscr{D}_{\text{th}}$ at which $\bar{g}' = 0$ is the limit of a sequence of points at which $\bar{g}' > 0$. At every point of such a sequence, $\bar{\Phi}$ determines the values of $\bar{\Lambda}_V$ and $\bar{K}_V$ by means of (14.6b). Since $\bar{\Lambda}_V$ and $\bar{K}_V$ are continuous, the limits of these sequences of values of $\bar{\Lambda}_V$ and $\bar{K}_V$ exist and determine the values of $\bar{\Lambda}_V$ and $\bar{K}_V$ at every point of $\mathscr{D}_{\text{th}}$ where $\bar{g}' = 0$. Thus Φ is a thermodynamic potential on $\mathscr{D}_{\text{th}}$.

In the older literature the relation $\Phi = \bar{\Phi}(V, \theta)$ was called a "fundamental equation" or "caloric equation of state", and the first equation of (14.6a) was interpreted as showing that a caloric equation of state determines the thermal equation of state (2.1). The converse is false: As is shown by the observation in the remark following Definition 19, the thermal equation of state restricts, but does not determine, a caloric equation of state.

Theorem 14B.   *Let $\mathscr{D}_{\text{th}}$ be such that $H = \bar{H}(V, \theta)$ is invertible for θ:*

$$\theta = \hat{\theta}(V, H). \tag{14.7}$$

If for any $\bar{f}$

$$\hat{f}(V, H) \equiv \bar{f}(V, \hat{\theta}(V, H)), \tag{14.8}$$

then

$$\hat{g} = \frac{\partial \hat{E}}{\partial H}, \qquad \hat{\varpi} = -\frac{\partial \hat{E}}{\partial V},$$

$$\hat{\Lambda}_V = -\hat{K}_V \frac{\partial \hat{\theta}}{\partial V}, \qquad \frac{\hat{g}}{\hat{h}} \hat{K}_V = \frac{\partial \hat{E}}{\partial H} \bigg/ \frac{\partial \hat{\theta}}{\partial H}. \tag{14.9}$$

Proof.   Since $\partial \bar{H}/\partial \theta > 0$ by the second equation of (9.26), the Jacobian determinant of the transformation from $\mathscr{D}_{\text{th}}$ in the V-θ plane to the V-H plane is always positive. The inverse function theorem delivers (14.7) locally, and the function $\hat{\theta}$ has continuous partial derivatives of second order whenever it exists. We derive (14.9) from the relations $\partial\theta/\partial V = -(\partial\bar{H}/\partial V)/(\partial\bar{H}/\partial\theta)$, $\partial\theta/\partial H = 1/(\partial\bar{H}/\partial\theta)$, (9.26), (9.27), and the chain rule. $\square$

Remark 1.   As has been noted in the proof, the inversion (14.7) is always possible locally in $\mathscr{D}_{\text{th}}$. If $\mathscr{D}_{\text{th}}$ is isochorically convex, the inversion (14.7) is possible throughout $\mathscr{D}_{\text{th}}$. All familiar examples of $\mathscr{D}$ are isochorically convex. Compare, *e.g.*, the constitutive domain of a Van der Waals fluid illustrated in Figure 6 in Chapter 5.

Remark 2.   Suppose the functions $\bar{g}$, $\bar{h}$, and $\hat{E}$ be known. Because $\bar{g}$ is an increasing function of the temperature, the first equation of (14.9) determines the function $\hat{\theta}$ in (14.7). The relations set forth in the second and fourth equations of (14.9) then determine $\bar{\varpi}$ and $\bar{K}_V$. Finally, $\bar{\Lambda}_V$ is determined from the third equation of (14.9). Thus $\hat{E}$ is a thermodynamic potential on $\mathscr{D}_{\text{th}}$.

Theorem 14C.   *Let $\mathscr{D}_{\text{th}}$ be such that (2.1) and $H = \bar{H}(V, \theta)$ can be solved to yield both V and θ as functions of p and H:*

$$V = \tilde{V}(p, H), \qquad \theta = \tilde{\theta}(p, H). \tag{14.10}$$

If

$$\tilde{f}(p, H) \equiv \bar{f}(\tilde{V}(p, H), \tilde{\theta}(p, H)), \tag{14.11}$$

then

$$\tilde{V} = \frac{\partial\tilde{X}}{\partial p}, \qquad \tilde{g} = \frac{\partial\tilde{X}}{\partial H}. \tag{14.12}$$

Proof.   It can be verified that the Jacobian determinant of the transformation from $\mathscr{D}_{\text{th}}$ in the V-θ plane to the p-H plane is negative. Hence it is always possible to solve locally for V and θ as in (14.10). Further, the functions $\tilde{V}$ and $\tilde{\theta}$ in (14.10) have continuous partial derivatives. By an application of the chain rule, (14.12) then follows. $\square$

Remark.   It is easy to verify that $\tilde{X}$ is a thermodynamic potential on $\mathscr{D}_{\text{th}}$. The proof of the next theorem and the verification that $\check{Z}$ is a thermodynamic potential will be left to the reader.

Theorem 14D.   *Let $\mathscr{D}_{\text{th}}$ be such that (2.1) is invertible for V:*

$$V = \check{V}(\theta, p). \tag{14.13}$$

If

$$\check{f}(\theta, p) \equiv \bar{f}(\check{V}(\theta, p), \theta), \tag{14.14}$$

then

$$\check{V} = \frac{\partial\check{Z}}{\partial p}, \qquad \check{g}'\check{H} = -\frac{\partial\check{Z}}{\partial\theta}. \tag{14.15}$$

Remark.   The inversion (14.13) is always possible locally in $\mathscr{D}_{\text{th}}$. A sufficient condition for the existence of $\check{V}$ throughout $\mathscr{D}_{\text{th}}$ is that $\mathscr{D}_{\text{th}}$ be isothermally convex. This condition fails for a Van der Waals fluid at sub-critical pressures, it being assumed that the entire constitutive domain of that fluid is a thermodynamic part. A glance at Figure 2 in Chapter 2 shows that for such a fluid there are two different functions $\check{V}$ in the subcritical range. One of these is appropriate to low reduced volumes and hence to the liquid phase; the other, to large reduced volumes and hence to the gaseous phase. Therefore at subcritical pressures there are *two distinct free enthalpies $\check{Z}$*, one for the liquid phase and another for the gaseous phase. Their domains of definition in the θ-p plane overlap.

Historical Scholion. REECH had only (8.18) at his disposal and not the Efficiency Theorem as expressed by (10.7). Thus results as explicit as those we have given here were not within his reach. Nevertheless, he introduced all four functions $\bar{\Phi}$, $\hat{E}$, $\bar{X}$, and $\check{Z}$ and proved them to be thermodynamic potentials. He did so in terms of a function apparently, though speciously, more general than $(H - H_0)g$, H_0 being a constant. Namely, using a general integrating factor f for Q as in Chapter 6, he in effect set

$$R(\theta, H) \equiv \Gamma(\theta) \int_{H_0}^{H} f(V^*(\theta, Y), \theta) dY, \qquad (14.16)$$

$V^*(\theta, H)$ being the function obtained by inverting $H = \bar{H}_f(V, \theta)$ for V, and Γ being the function that appears in (8.18). He nowhere perceived that f could be taken as a function of θ alone. In terms of his function R he defined functions apparently, though only apparently, more general than those we have called $\bar{\Phi}$, $\hat{E}$, $\bar{X}$, and $\check{Z}$, and he proved them to be thermodynamic potentials. In this sense we justly attribute to him the basic idea of the more specific results collected above in Theorems 14A, 14B, 14C, and 14D, and we regard those results collectively as expressing REECH's *Fourth Theorem*.

Nevertheless, having learnt that if a man allow an inch, however justly, to another, he stands to lose a yard of his own cloth, we remind the reader that so as to construct our theory we make no direct use of EULER's theorem on the existence of an integrating factor for a differential form in two variables. Our Theorem 8 in Chapter 9 shows that the specific function h, provided by Theorem 7, is an integrating factor over $\mathcal{D}_{\text{th}}$.

Corollary 14.1 (Reciprocal Relations, REECH, 1853).

$$\frac{\partial \bar{\varpi}}{\partial \theta} = \bar{g}' \frac{\partial \bar{H}}{\partial V},$$

$$\frac{\partial \hat{\varpi}}{\partial H} = -\frac{\partial \hat{g}}{\partial V},$$

$$\frac{\partial \bar{V}}{\partial H} = \frac{\partial \tilde{g}}{\partial p}, \qquad (14.17)$$

$$\frac{\partial \check{V}}{\partial \theta} = -\check{g}' \frac{\partial \check{H}}{\partial p}.$$

Remark. These relations are conditions of integrability for the functions $\bar{\Phi}$, $\hat{E}$, $\hat{X}$, and $\hat{Z}$, respectively. They are immediate consequences of (14.6a), the first and second equations of (14.9), (14.12), and (14.15), respectively. Conversely, if they are satisfied, then there are functions $\bar{\Phi}$, $\hat{E}$, $\hat{X}$, and $\hat{Z}$ such as to satisfy, respectively, (14.6a), the first and second equations of (14.9), (14.12), and (14.15).

Historical Comment. Relations of this kind, in appropriate special cases, go back to the work of CARNOT and are scattered through the classical papers. The tradition of thermodynamics seems to have adopted them first from a textbook by MAXWELL which appeared in 1871, nearly two decades after REECH had published them all in more general forms, forms general enough to include both CARNOT's thermodynamics and CLAUSIUS'. For one of many early examples, we may note that in the caloric theory the first equation of (14.17) is neither more nor less than the Carnot–Clapeyron Theorem (9.21), which played a central role in thermodynamics when CLAUSIUS was two years old and MAXWELL not yet born.

Corollary 14.2 (Differential Joule–Thomson Effect).   *If (2.1) is invertible for V, so that (14.13) holds, then along a curve in the θ-p plane the enthalpy $\hat{X}$ is constant if and only if*

$$\frac{\check{g}}{h} \check{K}_p \theta' + \left(\check{V} + \frac{\check{g}}{h} \check{\Lambda}_p \right) p' = 0, \tag{14.18}$$

the curve being given parametrically: θ = θ(s), p = p(s).

Proof. From (3.8) and the first equation of (9.28) it follows that $\check{\Lambda}_p = \check{h} \partial \check{H}/\partial p$ and $\check{K}_p = \check{h} \partial \check{H}/\partial \theta$. To obtain (14.18) we need only calculate $\partial \check{X}/\partial \theta$ and $\partial \check{X}/\partial p$ using (14.15) and the third equation of (14.1). □

Remark. For a body of ideal gas the inversion (14.13) is always possible, and if the constitutive domain of the gas is a thermodynamic part, use of the first equation of (3.9) and (11.1) reduces (14.18) to

$$\frac{\check{g}}{h} \check{K}_p \theta' + \check{V} \left(1 - \frac{\check{g}}{\theta \check{g}'} \right) p' = 0. \tag{14.19}$$

Hence, for such a body, the condition

$$\bar{g} = A\theta, \tag{14.20}$$

A being a positive constant, is necessary and sufficient that, in the image of $\mathcal{D}_{\text{th}}$ in the θ-p plane, the curves of constant enthalpy be isothermal segments. A change of temperature along a curve of constant enthalpy in the θ-p plane is called the *Joule–Thomson Effect.* The preceding analysis shows that *the condition (14.20) is necessary and sufficient that ideal gases be exempt* *from the Joule–Thomson Effect.* If (14.20) holds, then $\partial \bar{E}/\partial V = 0$, and if $\mathscr{D}_{\text{th}}$ is isothermally convex, $\bar{E}$ is a function of θ only.

Historical Comment. The absence of the Joule–Thomson Effect in ideal gases is sometimes regarded as providing decisive proof that the thermodynamics of CLAUSIUS is superior to the thermodynamics of CARNOT. That is only because no-one ever before analysed the effect in terms general enough to allow CARNOT's theory a chance to compete. It is easy to construct examples of bodies of ideal gas which are compatible not only with Axioms I–III but also with the caloric theory and for which $\bar{g}$ has the form (14.20). We therefore conclude that according to CARNOT's theory, too, it is possible that a body of ideal gas may pass through a porous plug isothermally if the difference of the pressures at the inlet and outlet is small.

We may look more closely into the matter by considering a body of Van der Waals fluid whose constitutive domain is a thermodynamic part and for which $\bar{g}$ has the form (14.20). For such a body $\check{g}\check{K}_p$ is positive, and (14.18) then shows that θ is a function of p along a curve of constant enthalpy. Further, from the first equation of (3.9), the first equation of (11.14), and (14.18) we find that along this curve $d\theta/dp \gtreqless 0$ if and only if $\theta \lesseqgtr 2a/(Rb)$, respectively, provided we introduce the usual approximating assumptions that the volume V of the body is large in comparison with b and that $R\theta V$ is large in comparison with a. *Thus for a Van der Waals fluid the Joule–Thomson inversion temperature $2a/(Rb)$ is independent of the function h.*

Indeed, the magnitude of the differential Joule–Thomson Effect depends upon h, but the existence and sign of it, under the usual approximating assumptions, follows just as well from CARNOT's theory as from CLAUSIUS', provided that (14.20) holds. If we recall that this choice of $\bar{g}$ is one of those that CARNOT himself considered, we see that *the Joule–Thomson Effect, in the gross terms in which it is usually considered, offers no secure basis for choice* between various possible thermodynamics. It can be explained by applying CLAUSIUS' thermodynamics but cannot serve as a test of it.

In another sense, however, the common opinion is just: CARNOT's theory merely allows ideal gases to be exempt from the Joule–Thomson Effect; CLAUSIUS' theory requires them to be.

PART III

UNIVERSAL EFFICIENCY OF ORDINARY CARNOT CYCLES

Universal Efficiency of Ordinary Carnot Cycles Compatible with the Existence of an Ideal Gas with Constant Specific Heats. Proof of the "First Law" and "Second Law" of Thermodynamics.

Till now we have dealt with the general structure based on our first three axioms alone. We shall impose as a further axiom the claim of CARNOT (*cf.* the General Scholion after Axiom III in Chapter 8): Any two bodies that absorb the same amount of heat while undergoing an ordinary Carnot cycle between fixed operating temperatures have the same motive power. By exploiting this axiom we shall in the end determine the universal efficiency of ordinary Carnot cycles for all bodies.

Axiom IV (Universal Efficiency).   *The function G in Axiom III is the restriction of a universal function $G_u(x, y, z)$ defined for all positive x, y, and z such that $x > y$.*

Clarification.   The function G in Axiom III is constitutive and can vary from body to body. Axiom IV, on the other hand, asserts that if $\mathscr{C}$ is an ordinary Carnot cycle having operating temperatures θ^+ and θ^-, undergone by *any* fluid body, then

$$L(\mathscr{C}) = G_u(\theta^+, \theta^-, C^+(\mathscr{C})) > 0, \qquad (15.1)$$

the function G_u being the same for all fluid bodies. This axiom merely asserts the existence of a universal function G_u without specifying its form. In Remark 2 after Corollary 11.2 of Chapter 10 we have exhibited two examples of fluid bodies compatible with Axioms I–III for which G has different forms. At this stage, we cannot determine whether these examples be compatible with Axiom IV or not. If we assume that the first example is compatible with Axiom IV, then the function G_u is determined as follows: $G_u(x, y, z) = (x - y)z/x^2$, and the second example is then not compatible with Axiom IV. It is therefore clear that Axiom IV restricts fluid bodies to a subclass of those compatible with Axioms I–III. Further, if a particular fluid body compatible with Axioms I–III has a function $G(x, y, z)$ which is defined for all positive x, y, and z such that $x > y$ and is also compatible with Axiom IV, then G_u is determined once and for all by this example. The next axiom asserts that one such fluid body is compatible with Axiom IV and thus leads to the determination of G_u.

Remark. An examination of the proof of the Third Principal Lemma in Chapter 9 shows that with the help of Axiom IV the lemma can be generalized as follows: $(\partial \varpi/\partial \theta)/\Lambda_V$ has the same value at the ordinary points of *all* fluid bodies, located on the same isotherm, and this property is shared also by $(\partial \Lambda_V/\partial \theta - \partial K_V/\partial V)/\Lambda_V$. If we now assume, in anticipation of the next axiom, that on every isotherm there is an ordinary point belonging to a fluid body compatible with Axioms I–IV, then we can define universal functions $f_u(\theta)$ and $\phi_u(\theta)$ for all temperatures (*cf.* (9.13)) as follows:

$f_u(\theta) \equiv$ the common value of $(\partial \varpi/\partial \theta)/\Lambda_V$ at all ordinary points of
   any fluid body on the isotherm with temperature θ,

$\phi_u(\theta) \equiv$ the common value of $(\partial \Lambda_V/\partial \theta - \partial K_V/\partial V)/\Lambda_V$ at all ordinary
   points of any fluid body on the isotherm with temperature θ. (15.2)

Clearly these functions are well defined and continuous. The functions f and ϕ in the proof of Theorem 7 in Chapter 9 are restrictions of f_u and ϕ_u to the interval $\mathscr{I}$ defined by $\mathscr{D}_{th}$. We can thus introduce the functions $g_u(\theta)$ and $h_u(\theta)$ for all positive θ by the following definitions, which are similar to (9.16):

$$h_u(\theta) \equiv \exp \int_{\theta_0}^{\theta} \phi_u(r)dr,$$

$$g_u'(\theta) \equiv h_u(\theta)f_u(\theta), \tag{15.3}$$

θ_0 being some fixed temperature. By arguments similar to those in the proof of Theorem 7 we show that g_u is a continuously differentiable increasing function while h_u is a continuously differentiable positive function. It is also clear that the basic constitutive restrictions (9.1) and (9.2) now hold in the thermodynamic parts of all fluid bodies if in them the constitutive functions g and h are replaced by the universal functions g_u and h_u, respectively. The same replacement can be made in all the results involving g and h derived in the preceding pages. Thus if g_u and h_u are known, the functions g and h for the thermodynamic parts of all fluid bodies are determined. It is also easy to verify that if $\bar{g}_u, \bar{h}_u$ is another universal replacement for g, h in (9.1) and (9.2), then

$$\bar{g}_u = Kg_u + N,$$

$$\bar{h}_u = Kh_u, \tag{15.4}$$

$K > 0$ and N being constants (*cf.* (9.17)). It follows from (10.7) that for all arguments x, y, z that correspond to an ordinary Carnot cycle in a thermo-dynamic part of any fluid body, the function G_u is given by

$$G_u(x, y, z) = \frac{g_u(x) - g_u(y)}{h_u(x)} z. \tag{15.5}$$

The next axiom enables us to determine g_u, h_u, and G_u once and for all.

Axiom V (Existence of an Ideal Gas Thermometer). *There is a body of ideal gas such that*

1. *the whole V-θ quadrant is a thermodynamic part; and*
2. *both specific heats are constant.*

Clarification. A body of ideal gas for which $\mathscr{D}$ = entire V-θ quadrant, $J\Lambda_V = R\theta/V$, and K_V = const. > 0 is easily seen to be compatible with Axioms I-III; also it satisfies all the conditions in Axiom V. Therefore, the real content of Axiom V lies in its assertion that one such body is also compatible with Axiom IV.

It should be unnecessary to point out that this axiom asserts something about the theory, not about nature. Nobody who understands classical mechanics will regard the theory of rigid bodies as implying that nature provides some bodies that are truly incapable of changing their shape, no matter what forces be applied to them. Rather, in many circumstances the changes of shape suffered by many natural bodies are so slight as to be negligible, so a conceptual model in which change of shape never occurs is both natural and useful. More specifically, a framework of classical mechanics that *disallowed* rigid bodies would not be sufficiently general. Axiom V meets a corresponding objection in classical thermodynamics. It asserts that *thermodynamics ought not forbid a mathematical model which represents the behavior of a gas as "ideal" and its specific heats constant*, no matter what be its volume and no matter what temperature it be subjected to.

General Scholion. To the modern student, accustomed to the example, this axiom seems feeble if not trivial. It is neither; rather, it is the key to classical thermodynamics. As TRUESDELL has shown in a historical lecture,[1] and as we have shown above in the Historical Scholion in Chapter 6, CARNOT's own theory contradicts it. See also Remark 4 after Theorem 15.

Theorem 15. *The universal functions g_u, h_u, and G_u have the following forms:*

$$g_u = Jh_u + \text{const.},$$

$$h_u = M\theta, \tag{15.6}$$

$$G_u(x, y, z) = J\left(1 - \frac{y}{x}\right)z,$$

J being a universal positive constant and M being an arbitrary positive constant.

1. *Op. cit.* Footnote 3 to Chapter 6.

Proof. The body of ideal gas whose existence is asserted by Axiom V will be used to determine the universal functions g_u, h_u, and G_u. From (11.3) we know that for this body $K_p > K_V$, and therefore we may define a positive constant J as follows:

$$J \equiv \frac{R}{K_p - K_V},\tag{15.7}$$

K_p and K_V being the constant specific heats of the body. From (3.12) we see that for this body Λ_V is given by

$$\Lambda_V = \frac{R\theta}{JV}.\tag{15.8}$$

Using (15.8) in (15.2), we show that $f_u = J/\theta$ and $\phi_u = 1/\theta$. The definitions (15.3) then lead to the first and second equations of (15.6). Since γ is constant for the body of ideal gas, a curve is an adiabat for it if and only if the LAPLACE-POISSON Law (4.10) holds on it. This fact when combined with (15.8) implies that for this body ordinary Carnot cycles operating between arbitrary temperatures and absorbing arbitrary amounts of heat exist. Substitution of the first and second equations of (15.6) into (15.5) then shows that G_u is given by the third equation of (15.6) for all positive x, y, and z such that $x > y$. □

Remark 1.   Axiom V can be replaced by other axioms which in conjunction with Axioms I–IV again lead to Theorem 15. One such replacement is the following axiom:

There is a body of ideal gas such that

i. *the whole V-θ quadrant is its constitutive domain;*
ii. *the body obeys* HOLTZMANN's *Assumption* (3.13) *or, equivalently,* MAYER's *Assumption* (3.14); *and*
iii. *the body interconverts heat and work uniformly in accord with Condition 3 of Corollary* 10.3 *in Chapter* 10.

The proof that this axiom again leads to Theorem 15 will be left to the reader. However, it seems more enlightening as well as more elegant to avoid assuming outright that heat and work are uniformly interconvertible even for a single body of ideal gas and instead to demonstrate uniform and universal interconvertibility of heat and work as a consequence of assumptions which seem weaker and are more specific and less sweeping. Axiom V is sufficient to achieve this goal. Other alternatives to Axiom V are discussed below and in Chapter 16.

The axiom just discussed is similar to one laid down by CLAUSIUS. The Appendix to this chapter presents a system of axioms which seems to express part of what CLAUSIUS assumed and does suffice to obtain CLAUSIUS' thermodynamics through simple, rigorous proofs.

We can obtain other alternatives to Axiom V by replacing the third condition in the axiom just discussed by *either* of the following two conditions:

iiiA. *γ is a function of θ alone.*
iiiB. *The body obeys the* LAPLACE-POISSON *Law* (4.10) *of adiabatic change.*

The proof that (iiiA) serves as a replacement is parallel to that of Theorem 15. To prove that (iiiB) serves as a replacement, we need only make use of the second result stated just after Property 9 in Chapter 11. We also note that (iiiA) itself can be replaced by any of the other five equivalent conditions listed in the remark after the second Historical Scholion in Chapter 11. Another replacement for (iiiA) would be: The speed of adiabatic sound is proportional to $\sqrt{\theta}$. Indeed, if such be the case, then Remark 2 after Corollary 4.1 in Chapter 4 shows that $\gamma = $ const.

Remark 2. As already noted in the remark after Axiom IV, in all our results involving g and h we may now replace g and h by g_u and h_u as given by the first and second equations of (15.6). The rest of this chapter is mainly a study of the consequence of making this replacement.

Remark 3. From the third equation of (15.6) we see that the efficiency of all ordinary Carnot cycles in every fluid body is given by the universal expression $1 - \theta^-/\theta^+$, θ^+ and θ^- being the operating temperatures of the cycle. As noted in the Scholion after Corollary 10.6 in Chapter 10, this classical expression obtained by the pioneers of thermodynamics is nowadays regarded as appropriate for "reversible" Carnot cycles.

Remark 4. In the Historical Scholion in Chapter 6 we have seen that *the caloric theory does not allow both specific heats of an ideal gas to be distinct constants.* CARNOT's results, of course, must deliver this conclusion *a fortiori.* The second example in Remark 2 after Corollary 11.2 of Chapter 10 shows that there are fluid bodies compatible with Axioms I–III and also with the caloric theory. Furthermore, if the example is compatible also with Axiom IV, then necessarily $G_u(x, y, z) = (x - y)z$. Thus it is clear that CARNOT's theory cannot be compatible with Axiom V, as is also evident from the second equation of (15.6). Indeed, the choice $h = 1$ reduces (11.3) and (11.4) to

$$K_p - K_V = \frac{R}{\theta g'},$$

$$K_V = -\frac{Rg''}{g'^2} \log V + K,$$

(15.9)

formulae due essentially to CARNOT. As we noted in the Historical Comment following Property 4 in Chapter 11, CARNOT himself considered the case in which $K_V = f(\theta)$, which is equivalent to $g' = $ const. If $K_V = $ const., we then see from the first equation of (15.9) that CARNOT's theory forces K_p to be a *decreasing function of θ.* This theoretical failure, more than any particular matter of experimental fact, doomed CARNOT's theory at birth.

The reader of this tractate will see in it detailed vindication of CLAUSIUS' judgment: CARNOT's ideas of the connection between work and heat were largely independent of the special and unacceptable theory of heat to which he applied them.

Remark 5. Although Axioms I–III by themselves do not imply that $\gamma = f(\theta)$ for an ideal gas, the adjunction of Axioms IV and V yields Theorem 15, so $g'_u = $ const., and (11.2) and the equation in (11.3) then show that $\partial\gamma/\partial V = 0$. If γ is constant on every adiabat, use of (4.3) shows that $\partial\gamma/\partial\theta = 0$, and therefore γ is constant. Making use of a result stated in Remark 3 after Corollary 4.3 in Chapter 4, we have a converse to the LAPLACE–POISSON Law: *Let the constitutive domain of a body of ideal gas be a thermodynamic part. If for such a body $\theta V^{\gamma-1} = $ const. along every adiabat, then $\gamma = $ const. for the body.*

The reader will recall that in the passage following Property 9 in Chapter 11 we have outlined a proof that (4.13) characterizes those ideal gases that satisfy both MAYER's Assumption and the LAPLACE-POISSON Law, within the theory provided by Axioms I–III only. When we adjoin Axioms IV and V, we obtain Theorem 15 and hence know that $h = M\theta$ and that all ideal gases obey MAYER's Assumption. In (4.13) we must therefore put $C = 0$. Thus we obtain a second proof of the converse to the LAPLACE-POISSON Law. This proof, strictly speaking, requires the assumption that $\mathscr{D}$ be isochorically convex. However, we see directly from (11.2) and (4.14) that $\partial K_V/\partial V = \partial K_V/\partial\theta = 0$. Therefore, the assumption that $\mathscr{D}$ be isochorically convex is unnecessary.

Corollary 15.1 (The "First Law of Thermodynamics"; Efficiency of C-Processes in a Thermodynamic Part). *Let $\mathscr{D}_{th}$ denote any thermodynamic part corresponding to some fluid body. For a cyclic process in $\mathscr{D}_{th}$ undergone by that fluid body*

$$L = JC, \tag{15.10}$$

J being the universal constant appearing in (15.6). Every body interconverts heat and work uniformly and universally while undergoing cyclic processes in $\mathscr{D}_{th}$. Each body has in $\mathscr{D}_{th}$ an internal energy $E(V, \theta)$ such that almost always

$$\dot{E} = JQ - \varpi\dot{V}; \tag{15.11}$$

also

$$J\Lambda_V - \varpi = \frac{\partial E}{\partial V}, \qquad JK_V = \frac{\partial E}{\partial\theta}. \tag{15.12}$$

For any body undergoing a C-process in $\mathscr{D}_{th}$

$$\frac{C^-}{C^+} = \frac{\theta^-}{\theta^+}, \qquad \frac{L}{JC^+} = 1 - \frac{\theta^-}{\theta^+}, \tag{15.13}$$

θ^+ being the absorption temperature and θ^- the emission temperature. If a heat-absorbing process in $\mathscr{D}_{th}$ that has distinct greatest and least temperatures θ_{max} and θ_{min} is not a Carnot process with these operating temperatures, its efficiency is less than $1 - \theta_{min}/\theta_{max}$.

Proof. Using (15.6) in Corollaries 10.3, 10.4, and 10.6 of Chapter 10, we arrive at (15.10) and (15.13). To obtain (15.11) and (15.12), we need only set $E \equiv E_{JM\theta, M\theta}$ in the second equation of (9.28) and in (9.27). The last conclusion is immediate from (13.6) and the condition for the attainment of the upper bound. □

Remark 1. The constant J is *the mechanical equivalent of a unit of heat.*

Remark 2. In Remark 3 after Theorem 15 we have seen that the classical expression for efficiency is valid for every ordinary Carnot cycle. From the second equation of (15.13) it follows that the classical result holds also for all C-processes in $\mathscr{D}_{\text{th}}$ and hence, in particular, for all Carnot cycles in $\mathscr{D}_{\text{th}}$.

Corollary 15.2 (CLAUSIUS' "Second Law of Thermodynamics"). *Each body has in $\mathscr{D}_{\text{th}}$ an entropy $H(V, \theta)$ such that*

$$Q = \theta \dot{H} \text{ almost always,} \qquad \Lambda_V = \theta \frac{\partial H}{\partial V}, \qquad K_V = \theta \frac{\partial H}{\partial \theta}. \quad (15.14)$$

Moreover,

$$\Lambda_V = \theta \left(\frac{\partial \Lambda_V}{\partial \theta} - \frac{\partial K_V}{\partial V} \right),$$

$$J\Lambda_V = \theta \frac{\partial \varpi}{\partial \theta}, \quad (15.15)$$

$$J(K_p - K_V) = -\theta \left(\frac{\partial \varpi}{\partial \theta} \right)^2 \Big/ \frac{\partial \varpi}{\partial V} \geqq 0.$$

Proof. From Theorem 15 it is clear that the choice $h = \theta$ and $g = J\theta$ is always possible in $\mathscr{D}_{\text{th}}$ for any body. If we now set $H \equiv H_\theta$, (15.14) follows by use of the first equation of (9.28) and of (9.26). To obtain (15.15) we need only make use of the above choice in (9.1), (9.2), and the second equation of (3.9). □

Remark 1. We leave it to the reader to notice that the choice $h = \theta$ and $g = J\theta$ reduces the results of Chapter 14 to the classical theorems on thermodynamic potentials. He will see at once also that a point is neutral if and only if it is piezotropic; that $\partial^2 \varpi / \partial V \partial \theta$ and $\partial^2 \varpi / \partial \theta^2$ exist and are continuous; and, by (9.19), that

$$\frac{J}{\theta} \frac{\partial K_V}{\partial V} = \frac{\partial^2 \varpi}{\partial \theta^2}. \quad (15.16)$$

These assertions, of course, refer to $\mathscr{D}_{\text{th}}$.

Remark 2 (Extension of Corollaries 15.1 and 15.2). Corollaries (15.1) and (15.2) were proved only for thermodynamic parts. Actually Axioms IV and V are strong enough to extend these results to more general parts of a constitutive domain. Substitution of the expressions for f_u and ϕ_u after (15.8) in (15.2) shows that the relations (15.15) hold at every ordinary point and by continuity also at *every* thermodynamic point in *all* fluid bodies. In particular, the first and second equations of (15.15) are satisfied in every set of thermodynamic points belonging to a constitutive domain whether or not that set be also a thermodynamic part. However, the point relations given in the first and second equations of (15.15) do not generally imply the existence of (single-valued) functions H and E satisfying (15.12) and the second and third equations of (15.14). If the set of thermodynamic points is also simply connected and open, such functions clearly exist. Thus Corollaries 15.1 and 15.2 remain true if in their statements $\mathscr{D}_{\text{th}}$ is replaced by any simply connected, open set of thermodynamic points in $\mathscr{D}$. Thus they apply to sets which do not satisfy the second condition for thermodynamic parts in Definition 17 of Chapter 9, *e.g.*, the constitutive domain in Figure 7A of Chapter 5. Likewise, the assertions of Remark 1 hold in any open, simply connected set of thermodynamic points.

Principal Scholion.   *We have shown that if Axioms* I *and* II *are accepted, then Axioms* III–V, *which refer to Carnot cycles and ideal gases, suffice to establish the formal structure of the thermodynamics of* CLAUSIUS *and* KELVIN, *thereby setting aside the theories of* LAPLACE, CARNOT, *and* CLAPEYRON. *The traditional "First Law" and "Second Law" of thermodynamics, as far as "reversible" processes in fluids are concerned, have been proved as consequences of those axioms.* They are included here as parts of Corollaries 15.1 and 15.2.

The "First Law" and the "Second Law", if put at the head of a development of thermodynamics, are general, sweeping statements. Experiments may conform to them; indeed, we know now that they do; but they are far too general to be *suggested* by experiment. The system of axioms we have set up and studied here refers *mainly to heat engines and to ideal gases*. Every one of its axioms is accessible in concept to a simple program of experiment. Statements (2), (ii), (iiiA), and (iiiB) in Axiom V and its alternative forms refer to specific properties of gases that were subjected to experiment in the early days of thermodynamics and were then found to hold fairly well for many real gases in a wide range of conditions.

A System of Axioms for the Thermodynamics of Clausius.

In his basic paper[1] on thermodynamics CLAUSIUS considered mainly ideal gases as far as his mathematical work was concerned, but he made a number of statements about fluid bodies in general. His starting point was Axioms I and II of this tractate; he assumed also the uniform and universal interconvertibility of heat and work in all cyclic processes:

Axiom IIIC. *For every cyclic process*
$$L = JC, \tag{15A.1}$$
the positive constant J being the same for all fluid bodies.

As we may see from Corollary 10.3 in Chapter 10, in a thermodynamic part this axiom is compatible with CARNOT's General Axiom. Of course it is a special instance of one of the statements now called "The First Law of Thermodynamics".

CLAUSIUS, indeed, presented an argument which he regarded as supporting a statement[2] equivalent to CARNOT's General Axiom and Axiom IV, but to

1. R. CLAUSIUS, "Über die bewegende Kraft der Wärme und die Gesetze, welche sich daraus für die Wärmelehre selbst ableiten lassen", *Annalen der Physik* (3) **19** = **79** (1850), 368–398, 500–524 = [with annotations] CLAUSIUS' *Abhandlungen* **1**, 16–78 (1864). Translation, "On the moving force of heat, and the laws regarding the nature of heat itself which are deducible therefrom", *Philosophical Magazine* (4) **2** (1851), 1–21, 102–119. A later English translation is conveniently available in MENDOZA's edition of CARNOT's treatise, New York, Dover Publications, 1960.

2. The statement was $L(\mathscr{C}) = H(\theta^+, \theta^-, C^-(\mathscr{C})) > 0$ for every ordinary Carnot cycle $\mathscr{C}$, the function H being universal; as we have noted in the second Historical Scholion of Chapter 11, CLAUSIUS made scant use of it. In assessing CLAUSIUS' dependence upon CARNOT's work we must recall that he knew it only through the inaccurate accounts of it given by CLAPEYRON and KELVIN.

us that argument seems to motivate two axioms, one of which is somewhat different:

Axiom IVC. *Every ordinary Carnot cycle does positive work.*

Axiom VC. *Let any two fluid bodies undergo ordinary Carnot cycles having common operating temperatures. If they do the same amount of work, they absorb the same amount of heat.*

Axioms I, II, and IIIC at once imply the existence of an internal energy E defined throughout $\mathscr{D}$ such that

$$J\Lambda_V - \varpi = \frac{\partial E}{\partial V}, \qquad JK_V = \frac{\partial E}{\partial \theta}. \qquad (15.12)_r$$

The First Law, in the form $\dot{E} = JQ - \varpi\dot{V}$ almost always, for every process, then follows. Also

$$\frac{\partial \varpi}{\partial \theta} = J\left(\frac{\partial \Lambda_V}{\partial \theta} - \frac{\partial K_V}{\partial V}\right) \qquad (10.19)_{2r}$$

at every point of $\mathscr{D}$. It should be noted that these results hold throughout $\mathscr{D}$, not merely in a thermodynamic part of it. Next, we construct two Carnot webs generated by ordinary Carnot cycles in neighborhoods of two ordinary points on the same isotherm, as indicated in Figure 14 of Chapter 9. Axiom IVC enables the selection of pairs of Carnot cycles, one from each web, for which Axiom VC applies. By constructing an appropriate sequence of such pairs we can show from Axiom VC that

$$\frac{\partial \varpi}{\partial \theta}\bigg/ \Lambda_V \text{ has the same (nonnegative) value} \qquad (15A.2)$$

at the ordinary points of all fluid bodies lying on a given isotherm (note the remark after Axiom IV).

CLAUSIUS himself, although he dismissed HOLTZMANN's work as pertaining only to the caloric theory, so as to complete his own deductions in effect adopted HOLTZMANN's Assumption:

Axiom VIC. *For an ideal gas*

$$J\Lambda_V = \varpi. \qquad (3.13)_r$$

Furthermore, he tacitly adopted another axiom:

Axiom VIIC. *There is a body of ideal gas having the entire V-θ quadrant as its constitutive domain.*

From these axioms and from (15A.2) it follows at once that *for all fluids*

$$\frac{\partial \varpi}{\partial \theta}\bigg/ \Lambda_V = J/\theta \qquad (15A.3)$$

at all ordinary points. Appeal to the continuity of $\partial \varpi / \partial \theta$ and Λ_V shows that

$$\theta \frac{\partial \varpi}{\partial \theta} = J\Lambda_V \qquad (15.15)_{2r}$$

at the thermodynamic points of all fluid bodies. For a simply connected, open set of thermodynamic points of a fluid body the existence of an entropy H satisfying (15.14) is then immediate. The foregoing arguments deliver an internal energy over all of $\mathscr{D}$ but an entropy only over simply connected, open subsets of it which consist of thermodynamic points only (note Remark 2 after Corollary 15.2).

From the second equation of (10.19) and the second equation of (15.15) we may derive various formulae of the classical thermodynamics of reversible processes in fluid bodies. As an example, Corollary 11.2 of Chapter 10 enables us to conclude that Axioms III and IV are satisfied, G_u having the form of the third equation of (15.6).

We have shown that a judicious selection of the various ideas CLAUSIUS expressed in his paper of 1850 would have provided him with a single set of axioms from which he could have obtained by clean mathematics, available in his day, not only the assertions and formulae of that paper concerning ideal gases but also their generalizations to all fluid bodies. However, the course of history was otherwise.

For bodies having ordinary points Axioms IIIC and IVC exclude the caloric theory from the start. The reader of this tractate will recall from Corollary 10.3 of Chapter 10 that within the framework of CARNOT's ideas Axiom IIIC in a thermodynamic part is equivalent to $g = Jh + $ const. Axiom VC has the effect of making h a universal function, while Axiom VIC evaluates h explicitly as $M\theta$.

Axioms I, II, and IIIC–VIIC provide a satisfactory basis for teaching classical thermodynamics efficiently and accurately to beginners who are not interested in comparing the theory accepted today with its predecessor or with other theories of the same kind.

CHAPTER 16

Invariance of the Carnot Function under Change of the Unit of Temperature. Alternative to Axiom V.

We may establish the results of Chapter 15 on the basis of two other axioms which, together, serve as an alternative to Axiom V. The first of these other axioms is

Axiom Vα. *The Carnot function G in Axiom* III *is invariant under change of the unit of temperature:*

$$G(x, y, z) = G(Kx, Ky, z) \qquad (16.1)$$

for all positive K such that both x, y, z and Kx, Ky, z are in the domain of G.

Remark 1. Since physical principles must be independent of the units in which quantities entering them are expressed, it might seem that Axiom Vα should be unnecessary. That is not true, even were we to regard the axioms of thermodynamics as mere additions to, or restrictions upon, some overriding system of axioms of physics. Firstly, in (16.1) the changes in the unit of temperature are restricted by the requirement that both x, y, z and Kx, Ky, z be in the domain of G. Even if arbitrary changes are allowed by assuming that $G(x, y, z)$ is defined for all positive x, y, and z such that $x > y$, the condition (16.1) does not necessarily follow from dimensional invariance. Since G is constitutive, it may well depend upon parameters characteristic of the particular body. One such parameter might be a temperature θ_0, e.g., the melting or boiling point of the body at some particular pressure. If G depends on such parameters having the dimension of temperature, the requirement of dimensional invariance clearly does not imply (16.1).

Remark 2. We may also impose on the universal function G_u of Axiom IV a requirement similar to that in Axiom Vα. Since $G_u(x, y, z)$ is defined for all positive x, y, and z such that $x > y$, the requirement is that

$$G_u(x, y, z) = G_u(Kx, Ky, z) \tag{16.2}$$

for all positive K. Now (16.2) does follow from dimensional invariance under change of the unit of temperature if it is assumed that G_u does not depend on any universal constant bearing the dimension of temperature.[1] From (16.2) we conclude at once that

$$G_u(x, y, z) = G_u\left(1, \frac{y}{x}, z\right). \tag{16.3}$$

The form for G_u obtained in Theorem 15 of Chapter 15 satisfies (16.2).

Remark 3. It is easy to exhibit examples of fluid bodies compatible with Axioms I–III that do or do not obey Axiom Vα. Both examples in Remark 2 following Corollary 11.2 of Chapter 10 fail to satisfy Axiom Vα, while the example in the Clarification after Axiom V in Chapter 15 does satisfy it.

Theorem 16. *Let only Axioms* I–III *and Axiom* Vα *be assumed. Then in* $\mathscr{D}_{\text{th}}$

$$h = M\theta^\delta, \qquad M = \text{const.} > 0, \qquad \delta = \text{const.} \tag{16.4}$$

If $\delta = 0$,

$$g = A \log \theta + \text{const.}, \tag{16.5}$$

A being a positive constant; if $\delta \neq 0$, *then*

$$g = Ah + \text{const.}, \tag{16.6}$$

A being a constant such that $A\delta > 0$.

> *Proof.* Let B belong to the temperature interval $\mathscr{I}$ of $\mathscr{D}_{\text{th}}$. Then we can construct a Carnot web $\mathscr{W}_{A,D}$ generated by an ordinary Carnot cycle in $\mathscr{D}_{\text{th}}$ such that B is in $]A, D[$. Let C be a number in $]A, B[$. It is easy to see that if x and y are in $[C, D]$, Kx and Ky are in $[A, D]$ for any K in $]0, 1]$ sufficiently close to 1. Applying (10.7) and (16.1) to the cycles of the web $\mathscr{W}_{A,D}$ operating between the temperatures x, y and Kx, Ky, and absorbing the same amount of heat, we show that if K is any number in $]0, 1]$ sufficiently close to 1, and $C \leqq y \leqq x \leqq D$, then
>
> $$\frac{g(Kx) - g(Ky)}{h(Kx)} = \frac{g(x) - g(y)}{h(x)}. \tag{16.7}$$
>
> If we now set $\bar{g}(x) \equiv g(Kx)$ and $\bar{h}(x) \equiv h(Kx)$, (16.7) reduces to

---

[1]. It is taken for granted here that the unit of temperature can be varied arbitrarily while holding fixed the units of heat, of work, and of any parameter on which G_u may depend.

(8.10). From the first equation of (8.12) and (8.14) we conclude that for x in $[C, D]$,

$$g(Kx) = A(K)g(x) + B(K),$$
$$h(Kx) = A(K)h(x),$$
(16.8)

the functions $A(K)$ and $B(K)$ replacing the constants in the first equation of (8.12) and (8.14). From the second equation of (16.8) it is clear that $A(K)$ is a differentiable function. Differentiating the second equation of (16.8) with respect to K and setting $K = 1$, we show that $xh'(x)/h(x) = $ const. on $[C, D]$. Hence the function $\theta h'/h$ is differentiable at B, and its derivative is 0. Since B is any temperature in $\mathscr{I}$, it follows that $\theta h'/h = \delta$ throughout $\mathscr{I}$, δ being a constant. Integration then leads to (16.4). Substitution of (16.4) in (16.8) yields

$$g(Kx) = K^\delta g(x) + B(K).$$
(16.9)

Again since $B(K)$ is differentiable, we differentiate (16.9) with respect to K and set $K = 1$ to obtain

$$xg'(x) = \delta g(x) + \text{const.},$$
(16.10)

for x in $[C, D]$. From (16.10) it is clear that $x^{\delta+1}(x^{-\delta}g)' = $ const. on $[C, D]$. Repeating the argument before (16.9), we show that $\theta^{\delta+1}(\theta^{-\delta}g)'$ is constant on $\mathscr{I}$, and integration then leads to (16.5) and (16.6). Since g is an increasing function, the constant A in (16.5) is positive, while the constant A in (16.6) is such that $A\delta > 0$. □

Remark 1.   The constant δ in (16.4) is clearly constitutive. We note also that $g' > 0$ always, irrespective of the value of δ. The following conclusions are immediate consequences of the above theorem and Corollaries 9.3, 10.3, and 10.4 in Chapter 10. For bodies that obey the caloric theory of heat $\delta = 0$, and g has the form (16.5) in every thermodynamic part $\mathscr{D}_{th}$. For bodies that interconvert heat and work uniformly, $\delta > 0$ in $\mathscr{D}_{th}$. For bodies that attain the classical efficiency in all ordinary Carnot cycles, $\delta = 1$ in $\mathscr{D}_{th}$. Conversely, in $\mathscr{D}_{th}$, if $\delta = 0$, $C^+ = C^-$ for every cyclic process; if $\delta > 0$, heat and work are uniformly interconvertible by cyclic processes; if $\delta = 1$, the classical efficiency is attained in all C-processes, and hence in every ordinary Carnot cycle.

Remark 2 (Necessary and sufficient condition for Axioms III and Vα). Corollary 11.2$_{ext}$ in Chapter 10 delivers a necessary and sufficient condition for Axiom III to hold. By using this condition and by slightly modifying the proof of Theorem 16 we may derive the following necessary and sufficient condition for Axioms III and Vα: *Let Axioms I and II be assumed. Then Axioms III and Vα are satisfied if and only if there are functions g and h on $\mathscr{O}$ that satisfy in each interval $\mathscr{I}_k$ all the requirements on them stated in Theorem 16 for $\mathscr{I}$ and are such that (9.1) and (9.2) hold at every ordinary point.* Theorem 16 is of course included as a special case of this result. It should be noted that the constants appearing in the expressions (16.4)–(16.6) for g and h *need not be the same* in each interval $\mathscr{I}_k$. Making use of the foregoing statement, we may easily construct examples of fluid bodies compatible with

Axioms I–III and Vα. By constructing a suitable example in this way, the reader may verify that the results of classical thermodynamics cannot be derived from these axioms alone.

Corollary 16.1 (Results for an Ideal Gas).   *For a body of ideal gas whose constitutive domain is a thermodynamic part, the difference of specific heats is a positive constant, and if a simple Carnot cycle $\mathscr{C}$ is labelled as in Figure* 11 *of Chapter* 7, *then*

$$\frac{\log \frac{V_c}{V_d}}{\log \frac{V_b}{V_a}} = \left(\frac{\theta^-}{\theta^+}\right)^{\delta-1}. \tag{16.11}$$

Further, the condition $V_b/V_a = V_c/V_d$ holds for every simple Carnot cycle if and only if it holds for one such cycle.

> *Proof.*   We arrive at these results by substituting in (11.3) and (11.7), the expressions for g and h in Theorem 16. The last conclusion is immediate from (16.11).   □

Remark.   For the body of ideal gas in the above corollary MAYER's Assumption (3.14) is satisfied for some positive constant J.

Axiom Vβ (Key Example).   *There is a body of ideal gas such that*

1. *the whole V-θ quadrant is a thermodynamic part; and*
2. *γ is a function of θ alone.*

Theorem 17.   *Let Axioms I–IV be satisfied. If Axioms Vα and Vβ replace Axiom V, the assertion of Theorem* 15 *still follows.*

> *Proof.*   By Corollary 16.1, MAYER's Assumption is satisfied. Therefore Axiom Vβ becomes equivalent to some of the alternatives to Axiom V discussed in Remark 1 after Theorem 15 in Chapter 15.   □

Remark.   In Remark 1 after Theorem 15 in Chapter 15 we have discussed various replacements for Condition (2) in Axiom Vβ. The last statement in Corollary 16.1 provides yet another replacement: $V_b/V_a = V_c/V_d$ for one simple Carnot cycle.

EPILOGUE

*Die Thermodynamik ist das Musterbeispiel einer
axiomatisch aufgebauten Wissenschaft.*
A. SOMMERFELD, 1952

CHAPTER 17

Axioms for Energy and Entropy.

The paper of REECH, extracted in 1851 and published in 1853, laid out the entire formal thermodynamics of "reversible" processes in homogeneous fluid bodies. Almost every equation ever obtained in that theory either is in REECH's paper or is an immediate corollary of equations in that paper. Nevertheless, the generality of REECH's results is in part illusory. Indeed, they do cover all of CARNOT's theory and all of CLAUSIUS', and every other possibility consistent with CARNOT's General Axiom, but also, since REECH did not obtain the central result given above as Theorem 10, the formulae he derived allow possibilities that CARNOT's General Axiom forbids.

CLAUSIUS, on the other hand, produced a theory so special as to exclude, from its very first step, most connections with CARNOT's specific results, although he did use what amounts to CARNOT's General Axiom along with other interdependent assumptions. Thence the reader of CLAUSIUS' papers, like the reader of a modern textbook, will easily regard CLAUSIUS' thermodynamics as being much more different from CARNOT's, or, to use a favorite adjective of historians of science, more "revolutionary", than in fact it is. CLAUSIUS' only important contribution to the thermodynamics of "reversible" processes, apart from his having made widely known some results REECH had already published, is his inference that (6.1) holds if f is taken as the temperature θ that occurs in the defining equation (2.9) of an ideal gas. Although no one will deny that CLAUSIUS somehow fell upon the right answer, his reasoning is mainly of the physical or metaphysical kind and does not carry conviction to a critical student.

Confining attention to thermodynamic parts, in the foregoing pages we have reduced REECH's results to the exact measure of his axioms, neither more general nor less so, as is confirmed by Theorem 11. In the first fourteen chapters of this tractate we have obtained and explicated the class of thermodynamic theories consistent with those axioms. We have proved that in all theories based on CARNOT's General Axiom, not merely in CLAUSIUS' special case, the heat-loss function h, which appears in the Efficiency Theorem, is an integrating factor for the heating. Thereupon, in Chapter 15, we have shown that if CARNOT's claim that the efficiency of an ordinary CARNOT cycle is a universal function be accepted, then CLAUSIUS' choice of θ for the integrating factor is justified as a mathematical consequence of allowing the possibility that an ideal gas may have constant specific heats. The "First Law" and the "Second Law" for "reversible" processes emerge as proved corollaries. The entire formal structure of the thermodynamics of "reversible" processes in fluid bodies follows by simple and explicit mathematics.

However, once that formal structure has been obtained, we easily see a shorter road to it. As REECH was the first to notice, use of a thermodynamic potential enables us to phrase everything in a few lines of formulae. Perhaps it was this fact that caused SOMMERFELD[1] to write

> In method there is a certain rivalry between *cyclic processes* and the method of *thermodynamic potentials*. The former are preferred for their immediacy, especially in engineering. We shall use the latter almost exclusively. They make possible a much shorter treatment, without any of the caprice attached to the artificially thought out cyclic processes.

Although we see nothing artificial or capricious in the treatment of cyclic processes presented in the foregoing pages—of course, a treatment which SOMMERFELD could not have seen—the brevity to which axiomatic introduction of a thermodynamic potential lends itself cannot be denied.

Another approach is through the axiomatic introduction of energy and entropy. The commonest and perhaps the most perspicuous formulation of this kind may be achieved by combining the assertions of Theorem 8 (Chapter 9) and Theorem 15 (Chapter 15). Thus we are led to set forth as primitive the following quantities associated with a fluid body at the time t:

Substate $\mathbf{Y}(t)$ [generalizing $V(t)$]
Temperature $\theta(t)$
Entropy $H(t)$
Heating $Q(t)$
Internal Energy $E(t)$
Net Working $W(t)$.

The substate $\mathbf{Y}(t)$ is a vector belonging to the real Euclidean space $\mathcal{R}^k$, variously identified, *e.g.*, as the volume of a mixture and as the masses of $k - 1$ of its k constituents. A process is then defined as a pair of piecewise

1. A. SOMMERFELD, Preface to *op. cit.* Footnote 4, Chapter 2.

smooth functions $\mathbf{Y}$, θ on an interval of time. As axioms we may lay down the following:

1. (First Law: Balance of Energy, Units Chosen so that $J = 1$):

$$\dot{E} = W + Q. \tag{17.1}$$

2. (Second Law: Balance of Entropy):

$$\theta\dot{H} = Q. \tag{17.2}$$

3. (First Constitutive Relation: Caloric Equations of State):

$$E = \bar{E}(\mathbf{Y}, \theta),$$
$$H = \bar{H}(\mathbf{Y}, \theta). \tag{17.3}$$

$\bar{E}$ and $\bar{H}$ are continuously differentiable on an open subset $\mathscr{D}$ of $\mathscr{R}^k \times$ $]0, \infty[$.

4. (Second Constitutive Relation: Linear Working):

$$W(t) = -\boldsymbol{\varpi}(\mathbf{Y}(t), \theta(t)) \cdot \dot{\mathbf{Y}}(t). \tag{17.4}$$

The function $\boldsymbol{\varpi}$, whose values are k-dimensional vectors, is a continuous function on $\mathscr{D}$.

5. (Compatibility): The functions $\bar{E}$, $\bar{H}$, and $\boldsymbol{\varpi}$ are such as to satisfy (17.1) and (17.2) in every process at each time when $\mathbf{Y}$ and θ are differentiable.

To see that these axioms do indeed lead to the formulae of classical thermodynamics, we need only eliminate Q between (17.1) and (17.2) and then substitute (17.3) and (17.4) into the results so as to obtain

$$\left(\frac{\partial\bar{E}}{\partial\mathbf{Y}} + \boldsymbol{\varpi} - \theta\frac{\partial\bar{H}}{\partial\mathbf{Y}}\right) \cdot \dot{\mathbf{Y}} + \left(\frac{\partial\bar{E}}{\partial\theta} - \theta\frac{\partial\bar{H}}{\partial\theta}\right)\dot{\theta} = 0. \tag{17.5}$$

Now (17.5) cannot hold for arbitrary processes $(\mathbf{Y}, \theta)$, as Axiom 5 requires it to do, unless

$$\boldsymbol{\varpi} + \frac{\partial\bar{E}}{\partial\mathbf{Y}} = \theta\frac{\partial\bar{H}}{\partial\mathbf{Y}}, \qquad \frac{\partial\bar{E}}{\partial\theta} = \theta\frac{\partial\bar{H}}{\partial\theta}. \tag{17.6}$$

Conversely, it is easy to see that Axioms 1, 3, and 4 together with this condition are also sufficient for Axiom 2 to hold in every process.

All the remaining formulae of classical thermodynamics follow effortlessly. For example, if in the special case in which $k = 1$ we use (15.12) so as to define Λ_V and K_V, we obtain at once the formal expression of the classical Doctrine of Latent and Specific Heats and, with some assumptions of smoothness, the thermodynamic formulae related to it, for example, (15.15) and the second and third equations of (15.14).

By itself, this rather abstract thermodynamics is scantly motivated. We present it here, much as TRUESDELL & TOUPIN in The Classical Field Theories[2] presented a parallel theory based on use of $\mathbf{Y}$ and H as independent variables, as a concise summary of a theory motivated elsewhere.

---

2. FLÜGGE's Handbuch der Physik III/1, 1960, Berlin etc., Springer-Verlag.

Nevertheless, even in the special case in which $k = 1$, the theory based on taking the properties of energy and entropy as axioms is in one sense a generalization, not merely a summary: it does not require any constitutive inequalities.

In our development we made use of the constitutive inequalities (2.2) and (3.2) to obtain some minor results, though the main results of this tractate have been established without the use of these inequalities. However, we did make essential use of the inequality in CARNOT's General Axiom (8.1). Axioms 1–5 do not require inequalities at all, for no analysis based upon them makes use of anything but differentiation. For example, we obtain right away CLAUSIUS' famous replacement for the Carnot–Clapeyron Theorem:

$$\Lambda_V = \theta \, \frac{\partial \varpi}{\partial \theta}. \qquad\qquad (15.15)_{2r}$$

From this formula we see that Λ_V and $\partial \varpi / \partial \theta$ must have the same sign and must vanish either together or never. Thus we prove, trivially, that a point of $\mathscr{D}$ is piezotropic if and only if it is neutral. This result is more specific than that given in Lemma 1 following CARNOT's General Axiom in Chapter 8, more specific even than that in Remark 1 following Corollary 15.2 in Chapter 15.

In the theory based on Carnot cycles as presented in this tractate the thermodynamic theorems have been proved only for thermodynamic parts or other special subsets of a constitutive domain. On the other hand, the theory based on axioms for energy and entropy leads effortlessly to results valid for the entire constitutive domain. For example, Axiom III can give us no information at all about $\partial \varpi / \partial \theta$ in a neutral part of $\mathscr{D}$, while the second equation of (15.15) proves at once that a neutral part is piezotropic. This appears to be one important difference between the two approaches. Another is that while the extension of the axiomatic structure to the case in which $k > 1$ is straightforward in the method based on energy and entropy, the same cannot be said for the notion of an ordinary Carnot cycle. In order to base a mathematical theory on the use of cycles of this kind, we must first define them precisely and then prove that they exist in sufficient abundance. This initial hurdle has to be crossed if Axiom III is to be retained without change when $k > 1$.

Scholion on the Teaching of Thermodynamics. The theory presented in the foregoing pages begins from simple and common caloric properties of fluid bodies and simple, immediate assumptions about cyclic engines in which fluids with those properties are the working substances. The internal energy and entropy are proved to exist, and in terms of them the entire theory takes the simple and explicit form developed, at last, in Chapter 15.

The student who learns thermodynamics in this way will be familiar with energy and entropy. He will easily see that in a theory intended to describe dissipation both Axiom I and Axiom II, which expresses the Doctrine of

Latent and Specific Heats, will have to be replaced by some more general class of constitutive relations. He will not be shocked if his teacher chooses to express the more general ideas of thermodynamics in terms not of calorimetry and heat engines but rather of energy and entropy, taken as primitive concepts and subjected to the general principles (17.1) and (17.2). In so doing he will follow the path that CLAUSIUS opened in 1854 though did not push far. He will follow also the tradition of mechanics and electromagnetism, in which proof that certain quantities exist in key special cases has been regarded as license to introduce those quantities thenceforth as abstract primitives so as to frame compact and comprehensive theories in terms of them.

To see how the theory of "irreversible" processes may be constructed as a simple generalization of the theory presented above, he may turn to Lecture I of TRUESDELL's *Rational Thermodynamics, A Course of Lectures on Selected Topics*, New York *etc.*, McGraw-Hill, 1969.

Index of Frequently Used Symbols

Symbol	Name	Place of Definition or First Major Occurrence
C	Net Gain of Heat	(1.1)
C^+	Heat Absorbed	(1.1)
C^-	Heat Emitted	(1.1)
$\mathscr{C}$	Cycle	Chapter 1, Def. 8
$\mathscr{D}$	Constitutive Domain	Chapter 1 Chapter 9,
$\mathscr{D}_n$	Normal Set of $\mathscr{D}$	Def. 17$_{bis}$
$\mathscr{D}_{th}$	Thermodynamic Part	Chapter 9, Def. 17
E	Internal Energy	(9.36)
$E_{g,h}$	Internal Pro-Energy	(9.27)
G	Carnot Function	(8.1)
g	Constitutive Function	(8.8)
$H \equiv H_\theta$	Entropy	Chapter 6
H_1	Heat Function	Chapter 6
H_f	Pro-Entropy (Corresponding to f)	(6.1)
H_h	Pro-Entropy	(9.26)
h	Heat-Loss Function	(8.8)
$\mathscr{I}$	Interval of Temperatures	Chapter 9
$\mathscr{I}_k$	Intervals in the Decomposition of $\mathcal{O}$	Chapter 9
J	Positive Constant Having the Dimension of Work/Heat	(3.13)
K_p	Specific Heat at Constant Pressure	(3.8)
K_V	Specific Heat at Constant Volume	(3.1)
L	Work	(1.6)
$\mathcal{O}$	Set of Temperatures corresponding to Ordinary Points $\mathscr{D}$	Chapter 9
$\mathscr{P}$	Path	Chapter 1, Def. 8
p	Pressure	(1.6)
Q	Heating	(1.1)
R	Constitutive Constant of a Body of Ideal Gas	(2.9)
$\mathscr{T}^+$	Set of Times During Which Heat is Absorbed	Chapter 1
$\mathscr{T}^-$	Set of Times During Which Heat is Emitted	Chapter 1
t	Time	Chapter 1
V	Volume	Chapter 1
$\mathscr{W}_{A,B}$	Carnot Web Whose Extreme Temperatures are B and A	Chapter 7, Def. 16
γ	Ratio of Specific Heats	(3.15)
θ	Temperature	Chapter 1
Λ_p	Latent Heat with Respect to Pressure	(3.8)
Λ_V	Latent Heat with Respect to Volume	(3.1)
ϖ	Pressure Function	(2.1)
ρ	Mass Density	Chapter 4
$\equiv$	Equality by Definition	Chapter 1

INDEX OF FREQUENTLY USED TERMS

Term	Place of Definition or First Occurrence
Absorption Temperature θ^+ of a C-Process	Chapter 10, Def. 18
Adiabat	Chapter 4, Def. 11
BRIDGMAN's "Hypothetical Liquid Water"	Chapter 2
Carnot Cycle, Process	Chapter 7, Def. 14
Carnot Function G	(8.1)
Carnot's General Axiom	Chapter 8
Carnot Web $\mathcal{W}_{A,B}$	Chapter 7, Def. 16
Constitutive Domain $\mathcal{D}$	Chapter 1
C-Process	Chapter 10, Def. 18
Cycle, Simple	Chapter 1, Def. 8
Doctrine of Latent and Specific Heats	Chapter 3
Efficiency	Chapter 8
Emission Temperature θ^- of a C-Process	Chapter 10, Def. 18
Energy, see "Internal Energy" and "Internal Pro-Energy"	
Entropy H_θ	Chapter 6
Extreme Temperatures B, A of a Web	Chapter 7, Def. 16
Heat Absorbed C^+, Heat Emitted C^-	Chapter 1, Def. 1
Heat Function H_1	Chapter 6
HOLTZMANN's Assumption	(3.13)
Ideal Gas	Chapter 2
Integrating Factor f	Chapter 6, Def. 13
Internal Energy E	(9.36)
Internal Pro-Energy $E_{g,h}$	Chapter 9
Isobar	Chapter 2, Def. 9
Isochor	Chapter 1
Isotherm	Chapter 1
MAYER's Assumption	(3.14)
Net Gain of Heat C	Chapter 1, Def. 1
Neutral Curve, Part, Point	Chapter 4, Def. 12
Normal Point, Set $\mathcal{D}_n$	Chapter 9, Def. 17_{bis}
Operating Temperatures θ^+, θ^- of a Carnot Process	Chapter 7, Def. 14
Ordinary Carnot Cycle	Chapter 7, Def. 15
Ordinary Point	Chapter 4, Def. 12
Path, Simple	Chapter 1, Def. 8
Piezotrope	Chapter 2, Def. 10
Piezotropic Part, Point	Chapter 2, Def. 10
Process	Chapter 1, Def. 2
Process, Adiabatic	Chapter 4, Def. 11
Process, Cyclic, Isothermal, Reverse of, Simple	Chapter 1
Pro-Energy, see "Internal Pro-Energy"	
Pro-Entropy H_f	Chapter 6, Def. 13
Specific heat K_p at Constant Pressure, K_V at constant volume	Chapter 3
Thermodynamic Part, Point $\mathcal{D}_{\text{th}}$	Chapter 9, Def. 17
Thermodynamic Potential	Chapter 14, Def. 19
Van der Waals Fluid	Chapter 2
Work	Chapter 1, Def. 3

Index of References to Historical Origins of Thermodynamics

Name	Page Nos.
CARNOT	5, 12, 22–24, 40, 41, 57, 58, 60, 66, 86, 88, 91, 94, 95, 101, 102, 106, 115, 116, 124, 129, 133, 147, 148
CLAPEYRON	19, 60, 66, 91, 116
CLAUSIUS	20, 21, 41, 58, 60, 75, 76, 90, 91, 106, 107, 115, 116, 132, 133, 137–139, 147, 148, 151
HOLTZMANN	24, 104, 138
JOULE	90
KELVIN	13, 20, 39, 55, 90, 91, 106, 116
LAPLACE	40
MAYER	24, 90, 104
POISSON	40
RANKINE	90, 91, 116
REECH	28, 38, 39, 41, 58, 65, 66, 75, 76, 87, 91, 119, 123, 124, 147, 148

Index of References to Standard Presentations of Thermodynamics

Name of Author	Page Nos.
EPSTEIN, P. S.	12
KESTIN, J.	18, 37, 74
PLANCK, M.	77
SOMMERFELD, A.	12, 49, 148